REVIEW OF THE NARSTO DRAFT REPORT

NARSTO ASSESSMENT OF THE ATMOSPHERIC SCIENCE ON PARTICULATE MATTER

Committee to Review NARSTO's Scientific Assessment of
Airborne Particulate Matter

NATIONAL RESEARCH COUNCIL
OF THE NATIONAL ACADEMIES

In collaboration with
Royal Society of Canada
United States–Mexico Foundation for Science
(FUMEC)

THE NATIONAL ACADEMIES PRESS
Washington, D.C.
www.nap.edu

THE NATIONAL ACADEMIES PRESS   500 Fifth Street, N.W.   Washington, D.C. 20001

NOTICE: The project that is the subject of this report was approved by the Governing Board of the National Research Council, whose members are drawn from the councils of the National Academy of Sciences, the National Academy of Engineering, and the Institute of Medicine. The members of the committee responsible for the report were chosen for their special competences and with regard for appropriate balance.

This project was supported by Contract No. X-825288-01-01 between the National Academy of Sciences and the U.S. Environmental Protection Agency. Any opinions, findings, conclusions, or recommendations expressed in this publication are those of the author(s) and do not necessarily reflect the view of the organizations or agencies that provided support for this project.

International Standard Book Number 0-309-08603-5

Additional copies of this report are available from:

The National Academies Press
500 Fifth Street, N.W.
Box 285
Washington, DC 20055

800-624-6242
202-334-3313 (in the Washington metropolitan area)
http://www.nap.edu

Copyright 2002 by the National Academy of Sciences. All rights reserved.

Printed in the United States of America

THE NATIONAL ACADEMIES
Advisers to the Nation on Science, Engineering, and Medicine

The **National Academy of Sciences** is a private, nonprofit, self-perpetuating society of distinguished scholars engaged in scientific and engineering research, dedicated to the furtherance of science and technology and to their use for the general welfare. Upon the authority of the charter granted to it by the Congress in 1863, the Academy has a mandate that requires it to advise the federal government on scientific and technical matters. Dr. Bruce M. Alberts is president of the National Academy of Sciences.

The **National Academy of Engineering** was established in 1964, under the charter of the National Academy of Sciences, as a parallel organization of outstanding engineers. It is autonomous in its administration and in the selection of its members, sharing with the National Academy of Sciences the responsibility for advising the federal government. The National Academy of Engineering also sponsors engineering programs aimed at meeting national needs, encourages education and research, and recognizes the superior achievements of engineers. Dr. Wm. A. Wulf is president of the National Academy of Engineering.

The **Institute of Medicine** was established in 1970 by the National Academy of Sciences to secure the services of eminent members of appropriate professions in the examination of policy matters pertaining to the health of the public. The Institute acts under the responsibility given to the National Academy of Sciences by its congressional charter to be an adviser to the federal government and, upon its own initiative, to identify issues of medical care, research, and education. Dr. Harvey V. Fineberg is president of the Institute of Medicine.

The **National Research Council** was organized by the National Academy of Sciences in 1916 to associate the broad community of science and technology with the Academy's purposes of furthering knowledge and advising the federal government. Functioning in accordance with general policies determined by the Academy, the Council has become the principal operating agency of both the National Academy of Sciences and the National Academy of Engineering in providing services to the government, the public, and the scientific and engineering communities. The Council is administered jointly by both Academies and the Institute of Medicine. Dr. Bruce M. Alberts and Dr. Wm. A. Wulf are chair and vice chair, respectively, of the National Research Council

www.national-academies.org

COMMITTEE TO REVIEW NARSTO'S SCIENTIFIC ASSESSMENT OF AIRBORNE PARTICULATE MATTER

Members

JOE L. MAUDERLY *(Chair)*, Lovelace Respiratory Research Institute, Albuquerque, New Mexico
MICHAEL BRAUER, The University of British Columbia, Vancouver, British Columbia, Canada
MAURICIO HERNANDEZ-AVILA, National Institute of Public Health, Cuernavaca, Morelos, Mexico
NANCY KETE, World Resources Institute, Washington, DC
CHARLES E. KOLB, Aerodyne Research, Inc., Billerica, Massachusetts
WILLIAM LEISS, R. Samuel McLaughlin Centre for Population Health, Ottawa, Ontario, Canada
GERARDO MANUEL MEJIA-VELAZQUEZ, Instituto Tecnologico y de Estudios Superiores de Monterrey, Monterrey, Nuevo Leon, Mexico
LUISA T. MOLINA, Massachusetts Institute of Technology, Cambridge, Massachusetts
LYNN RUSSELL, Princeton University, Princeton, New Jersey

Project Staff

K. JOHN HOLMES, Senior Staff Officer
RAYMOND WASSEL, Senior Program Director
LAURIE GELLER, Staff Officer
AMANDA STAUDT, Postdoctoral Research Associate
NORMAN GROSBLATT, Senior Editor
MIRSADA KARALIC-LONCAREVIC, Research Assistant
RAMYA CHARI, Research Assistant
EMILY SMAIL, Senior Project Assistant
KELLY CLARK, Editorial Assistant

BOARD ON ENVIRONMENTAL STUDIES AND TOXICOLOGY[1]

Members

GORDON ORIANS *(Chair)*, University of Washington, Seattle
JOHN DOULL *(Vice Chair)*, University of Kansas Medical Center, Kansas City
DAVID ALLEN, University of Texas, Austin
THOMAS BURKE, Johns Hopkins University, Baltimore, Maryland
JUDITH C. CHOW, Desert Research Institute, Reno, Nevada
CHRISTOPHER B. FIELD, Carnegie Institute of Washington, Stanford, California
WILLIAM H. GLAZE, University of North Carolina, Chapel Hill
SHERRI W. GOODMAN, Center for Naval Analyses, Alexandria, Virginia
DANIEL S. GREENBAUM, Health Effects Institute, Cambridge, Massachusetts
ROGENE HENDERSON, Lovelace Respiratory Research Institute, Albuquerque, New Mexico
CAROL HENRY, American Chemistry Council, Arlington, Virginia
ROBERT HUGGETT, Michigan State University, East Lansing
BARRY L. JOHNSON Emory University, Atlanta, Georgia
JAMES H. JOHNSON, Howard University, Washington, D.C.
JAMES A. MACMAHON, Utah State University, Logan
PATRICK V. O'BRIEN, Chevron Research and Technology, Richmond, California
DOROTHY E. PATTON, International Life Sciences Institute, Washington, D.C.
ANN POWERS, Pace University School of Law, White Plains, New York
LOUISE M. RYAN, Harvard University, Boston, Massachusetts
JONATHAN M. SAMET, Johns Hopkins University, Baltimore, Maryland
KIRK SMITH, University of California, Berkeley
LISA SPEER, Natural Resources Defense Council, New York, New York
G. DAVID TILMAN, University of Minnesota, St. Paul
LAUREN A. ZEISE, California Environmental Protection Agency, Oakland

Senior Staff

JAMES J. REISA, Director
DAVID J. POLICANSKY, Associate Director and Senior Program Director for Applied Ecology
RAYMOND A. WASSEL, Senior Program Director for Environmental Sciences and Engineering
KULBIR BAKSHI, Program Director for the Committee on Toxicology
ROBERTA M. WEDGE, Program Director for Risk Analysis
K. JOHN HOLMES, Senior Staff Officer
SUSAN N.J. MARTEL, Senior Staff Officer
SUZANNE VAN DRUNICK, Senior Staff Officer
RUTH E. CROSSGROVE, Managing Editor

---

[1]This study was planned, overseen, and supported by the Board on Environmental Studies and Toxicology.

BOARD ON ATMOSPHERIC SCIENCES AND CLIMATE

ERIC J. BARRON (*Chair*), Pennsylvania State University, University Park, Pennsylvania
RAYMOND J. BAN, The Weather Channel, Inc., Atlanta, Georgia
ROBERT C. BEARDSLEY, Woods Hole Oceanographic Institution, Woods Hole, Massachusetts
ROSINA M. BIERBAUM, The University of Michigan, Ann Arbor, Michigan
HOWARD B. BLUESTEIN, University of Oklahoma, Norman, Oklahoma
RAFAEL L. BRAS, Massachusetts Institute of Technology, Cambridge. Massachusetts
STEVEN F. CLIFFORD, National Oceanic and Atmospheric Administration, Boulder, Colorado
CASSANDRA G. FESEN, University of Texas at Dallas, Texas
GEORGE L. FREDERICK, Vaisala Meteorological Systems, Inc., Boulder, Colorado
JUDITH L. LEAN, Naval Research Laboratory, Washington, DC
MARGARET A. LEMONE, National Center for Atmospheric Research, Boulder, Colorado
MARIO J. MOLINA, Massachusetts Institute of Technology, Cambridge, Massachusetts
MICHAEL J. PRATHER, University of California, Irvine, California
WILLIAM J. RANDEL, National Center for Atmospheric Research, Boulder, Colorado
RICHARD D. ROSEN, Atmospheric & Environmental Research, Inc., Lexington, Massachusetts
THOMAS F. TASCIONE, Sterling Software, Inc., Bellevue, Nebraska
JOHN C. WYNGAARD, Pennsylvania State University, University Park, Pennsylvania

OTHER REPORTS OF THE
BOARD ON ENVIRONMENTAL STUDIES AND TOXICOLOGY

The Airliner Cabin Environment and Health of Passengers and Crew (2002)
Arsenic in Drinking Water: 2001 Update (2001)
Evaluating Vehicle Emissions Inspection and Maintenance Programs (2001)
Compensating for Wetland Losses Under the Clean Water Act (2001)
A Risk-Management Strategy for PCB-Contaminated Sediments (2001)
Toxicological Effects of Methylmercury (2000)
Strengthening Science at the U.S. Environmental Protection Agency: Research-Management and Peer-Review Practices (2000)
Scientific Frontiers in Developmental Toxicology and Risk Assessment (2000)
Copper in Drinking Water (2000)
Ecological Indicators for the Nation (2000)
Waste Incineration and Public Health (1999)
Hormonally Active Agents in the Environment (1999)
Research Priorities for Airborne Particulate Matter (three reports, 1998-2001)
Ozone-Forming Potential of Reformulated Gasoline (1999)
Arsenic in Drinking Water (1999)
Brucellosis in the Greater Yellowstone Area (1998)
The National Research Council's Committee on Toxicology: The First 50 Years (1997)
Carcinogens and Anticarcinogens in the Human Diet (1996)
Upstream: Salmon and Society in the Pacific Northwest (1996)
Science and the Endangered Species Act (1995)
Wetlands: Characteristics and Boundaries (1995)
Biologic Markers (five reports, 1989-1995)
Review of EPA's Environmental Monitoring and Assessment Program (three reports, 1994-1995)
Science and Judgment in Risk Assessment (1994)
Pesticides in the Diets of Infants and Children (1993)
Protecting Visibility in National Parks and Wilderness Areas (1993)
Dolphins and the Tuna Industry (1992)
Science and the National Parks (1992)
Assessment of the U.S. Outer Continental Shelf Environmental Studies Program, Volumes I-IV (1991-1993)
Human Exposure Assessment for Airborne Pollutants (1991)
Rethinking the Ozone Problem in Urban and Regional Air Pollution (1991)
Decline of the Sea Turtles (1990)

Copies of these reports may be ordered from the National Academy Press
(800) 624-6242 or (202) 334-3313
www.nap.edu

PREFACE

NARSTO[2] is a public-private partnership with members from government, utilities, industry, and academe in Canada, Mexico, and the United States. It was chartered in 1995 with a primary mission to coordinate and enhance policy-relevant atmospheric-science research on and assessment of tropospheric ozone. The scope of NARSTO's activities was expanded in 1998 to include airborne particulate matter (PM). Shortly after it was founded, NARSTO asked the National Research Council to establish a committee that could offer guidance on scientific questions, integration and assessment, short- and long-term balance issues, and research priorities and that could review NARSTO's program activities, progress, and draft products. The NRC Committee to Assess the North American Research Strategy for Tropospheric Ozone Program was convened in 1997 and reviewed NARSTO's first assessment, *An Assessment of Tropospheric Ozone Pollution–A North American Perspective*, released in 2000.

After expanding its scope, NARSTO prepared a draft of *NARSTO Assessment of the Atmospheric Science on Particulate Matter*. The assessment is intended to be a concise, scientifically credible, comprehensive discussion of atmospheric-science issues associated with managing ambient PM concentrations to meet adopted or proposed air-quality standards in the three North American nations. The assessment strives to provide policy-relevant scientific information without making specific policy recommendations. NARSTO requested that the NRC review the draft assessment in consultation with the Royal Society of Canada (RSC) and the United States-Mexico Foundation for Science (Fundacion Mexico Estados Unidos para la Ciencia, FUMEC). In response to that request, the NRC, in collaboration with RSC and FUMEC, created the Committee to Review NARSTO's Assessment of Airborne Particulate Matter. A framework for collaboration throughout the review of the NARSTO assessment is described in a memorandum of understanding that was approved by the three organizations.

---

[2]Originally, NARSTO stood for the North American Research Strategy for Tropospheric Ozone; however, when its charter was expanded to include PM, the term NARSTO became simply a word signifying the partnership.

This report presents the committee's consensus comments on the draft NARSTO assessment document dated December 31, 2001. Members of the public who wish to examine the draft assessment that was submitted to the committee should contact the National Academies Public Access Records Office at publicac@nas.edu.

Several people assisted the committee by providing information related to issues addressed in this report. I gratefully acknowledge Jeremy Hales, NARSTO management coordinator; Jeffrey West, NARSTO associate management coordinator; Marjorie Shepard, Meteorological Service of Canada, NARSTO assessment cochair; James Vickery, U.S. Environmental Protection Agency, NARSTO assessment cochair; Peter McMurry, University of Minnesota, NARSTO assessment cochair; George Hidy, Envair/Aerochem; Geoff Flynn, RSC; and Guillermo Fernandez de al Garza, FUMEC.

This report was reviewed in draft form by persons chosen for their diverse perspectives and technical expertise in accordance with procedures approved by the NRC's Report Review Committee. The purposes of the independent review were to provide candid and critical comments that assist the NRC in making the final report as sound as possible and to ensure that the report meets institutional standards for objectivity, evidence, and responsiveness to the study charge. The review comments and draft report of the committee remain confidential to protect the integrity of the deliberative process. We wish to thank the following for their review of the report: Margarita Castillejos, Metropolitan Autonomous University; Susanne Hering, Aerosol Dynamics Inc.; Steve Hrudey, University of Alberta; John McConnell, York University; Paulette Middleton, RAND; Luis Gerardo Ruiz Suarez, National Autonomous University of Mexico; Milton Russell, Joint Institute for Energy and Environment; Sergio Sanchez, Ministry of Natural Resources and Environment; and Andrew Sessler, E.O. Lawrence Berkeley National Laboratory.

Although the reviewers listed above have provided many constructive comments and suggestions, they were not asked to endorse the conclusions or recommendations, nor did they see the final draft of the report before its release. The review of this report was overseen by Richard Goody, Harvard University. Appointed by the National Research Council, he was responsible for making certain that an independent examination of the report was carried out in accordance with institutional procedures and that all review comments were carefully considered. Responsibility for the final content of this report rests entirely with the committee and the institution.

I am also grateful for the assistance of the National Research Council staff in preparing this report. Staff members who contributed to the effort include K. John Holmes, project director; Raymond Wassel, senior program director for environmental sciences and engineering in the Board on Environmental Studies and Toxicology (BEST); Warren Muir, executive director of the Division on Earth and Life Studies; James Reisa, director of BEST; Amanda Staudt, postdoctoral research associate; Laurie Geller, program officer in the Board on Atmospheric Science and Climate; Norman Grossblatt, senior editor; Mirsada Karalic-Loncarevic, research assistant; Ramya Chari, research assistant; and Emily Smail, senior project assistant.

Finally, I would like to thank all the members of the committee for their expertise and dedicated effort throughout the development of this report.

                                         Joe L. Mauderly, DVM
                                         Chair, Committee to Review NARSTO's Scientific
                                         Assessment of Airborne Particulate Matter

CONTENTS

SUMMARY 1

1 INTRODUCTION 6

2 OVERARCHING ISSUES 11
    PM Conceptual Model and Framework for Informing Airborne-PM Management, 11
    Other Content Issues, 13
    Presentation Issues, 15

3 COMMENTS ON THE ASSESSMENT'S EXECUTIVE SUMMARY 17
    Restructuring of the Executive Summary, 17
    Other Comments on the Executive Summary, 18
    Comments on the Policy Questions, 22

4 COMMENTS ON THE ASSESSMENT'S CHAPTERS AND RELATED
   APPENDIXES 31
    Chapter 1: "Perspective and Context for Managing PM", 31
    Chapter 2: "Atmospheric Aerosol Processes: How Particles Change While Suspended
       in the Air", 32
    Chapter 3: "Emission Inventories"; Appendix A: "Emission Calculations and
       Inventory Listings", 33
    Chapter 4: "Gas and Particle Measurements"; Appendix B: "Measurements", 34
    Chapter 5: "Spatial and Temporal Characterization of Particulate Matter Concentration
       and Composition"; Appendix C: "Monitoring Data: Availability, Limitations, and
       Network Issues"; Appendix D: "Global Aerosol Transport", 36
    Chapter 6: "Receptor Methods for Source Apportionment—Beyond the Emission
       Inventories", 39
    Chapter 7: "Using Models to Estimate Particle Concentrations and Exposure", 41

Chapter 8: "Health Effects of Particulate Matter", 43
Chapter 9: "Visibility Effects", 44
Chapter 10: "Conceptual Descriptions of PM for Nine North American Regions"; Appendix E: "Conceptual Descriptions of Selected North American Sites", 46
Chapter 11: "Recommendations", 48

5 RECOMMENDATIONS FOR FUTURE NARSTO ASSESSMENTS 51

REFERENCES 53

ATTACHMENT A. BIOGRAPHICAL INFORMATION ON THE COMMITTEE TO REVIEW NARSTO'S SCIENTIFIC ASSESSMENT OF AIRBORNE PARTICULATE MATTER 57

ATTACHMENT B. LINE-BY-LINE COMMENTS 60

Review of the NARSTO Draft Report: NARSTO Assessment of the Atmospheric Science on Particulate Matter

SUMMARY

Understanding the processes responsible for high concentrations of particulate matter (PM)[1] in the atmosphere is critical for air-quality managers. NARSTO[2], a public-private partnership with members from government, utilities, industry, and academe in Canada, Mexico, and the United States, was founded in 1995 to coordinate and enhance policy-relevant atmospheric science research and assessment on tropospheric ozone. The scope of NARSTO was expanded to include ambient PM in 1998. To inform policy-makers on how emissions of PM and its gaseous precursors are related to the distribution of ambient PM, NARSTO has recently completed a draft of the *NARSTO Assessment of the Atmospheric Science on Particulate Matter*. The assessment is intended to be a concise, scientifically credible, comprehensive discussion of the atmospheric-science issues associated with management of airborne PM to achieve air quality standards. The assessment strives to present the issues from a policy-relevant perspective and to develop a common scientific understanding that can be used in implementing new PM ambient air-quality standards.

The National Research Council (NRC), in consultation with the Royal Society of Canada (RSC) and the United States-Mexico Foundation for Science (Fundacion Mexico Estados Unidos para la Ciencia, FUMEC), was asked to review the draft NARSTO assessment, dated December 31, 2001. In response to that request, the NRC established the trinational Committee to Review NARSTO's Assessment of Airborne Particulate Matter. The committee was formed with input and concurrence from the RSC and FUMEC in accordance with a memorandum of understanding that was approved by the two organizations and the NRC.[3] In conducting its review, the committee was tasked to consider the stated overall goals and objectives of the assessment, the

---

[1] Airborne particulate matter refers to a broad class of discrete solid particles and liquid droplets of varied chemical composition and size. PM_{10} is defined as the mass of PM collected by a sampler with a 50% size cutoff at 10 µm in aerodynamic diameter; $PM_{2.5}$ is the mass of PM collected by a sampler with a 50% size cutoff at 2.5 µm.

[2] Originally, NARSTO stood for the North American Research Strategy for Tropospheric Ozone. When its charter was expanded to include PM, the term NARSTO became simply a word signifying the partnership.

[3] The review was conducted in accordance with standard NRC policies, including compliance with Section 15 of the U.S. Federal Advisory Committee Act.

scientific and technical analysis provided, and how effectively policy-relevant questions were addressed.

The draft NARSTO assessment consists of a comprehensive review of the science pertaining to PM and an executive summary intended to synthesize the scientific information for a decision-maker audience. Within the body of the assessment are technical chapters addressing formation and transport of PM, emission inventories, measurement techniques, ambient PM concentrations, source apportionment and air quality-models, health effects, and visibility. Some analysis of the technical information is provided via recommendations for future research needs and "conceptual descriptions" of the factors that affect observed ambient PM concentrations in nine regions in North America. The executive summary of the assessment contains a summary of the information presented in the body of the document and responses to eight policy questions, which are used to synthesize the atmospheric-science information necessary for implementing ambient PM standards. The assessment cochairs informed the committee that the executive summary will be published with the full document and also separately in English, Spanish, and French. Here, the committee summarizes the highest-priority recommendations for the chapters of the draft assessment and its executive summary.

COMMENTS ON THE CHAPTERS OF THE DRAFT NARSTO ASSESSMENT

The committee commends NARSTO for undertaking this assessment, particularly in light of the challenge it posed. The committee finds that the draft NARSTO PM assessment is a good representation of the state of atmospheric science, has an appropriate level of detail in most chapters, and provides the best currently available information on the formation and distribution of PM for North America. In particular, the draft clearly identifies the widespread existence and variable nature of the PM problem in different regions of North America. The draft provides an unmatched compilation of chemical measurements of both inorganic and organic fractions of PM, with preliminary indications of the sources and fates of the components. A comprehensive discussion of computational models shows how the weight of evidence defines a clear path for linking emission sources with air-quality outcomes. The committee finds the descriptions of PM characteristics and source contributions in nine regions to constitute an original scientific contribution and to provide information of great use to decision-makers. With suitable revisions, the document will be useful to the air-quality management community, to atmospheric-science researchers, and potentially to researchers in related fields. Indeed, the committee is unaware of any other assessments of the atmospheric science of PM that are this comprehensive; this NARSTO assessment is potentially of great use.

The draft assessment in its current form needs substantial improvements to meet its goal of providing useful information in a way that is accessible to decision-makers, especially in terms of consistently presenting a framework for informing airborne-PM management and stating clearly the policy implications of scientific findings. The committee's highest-priority recommendations for improving the main body of the assessment follow. Additional comments are provided in the chapters of the present report and its Attachment B.

- A framework for informing airborne-PM management, which includes a general conceptual model for understanding PM formation, is not explicitly presented and consistently used in the draft assessment. In Chapter 2 of this review, the committee offers a diagram that illustrates such a framework. The committee recommends that a framework of this sort be explicitly defined and introduced in Chapter 1 of the assessment and used consistently throughout it. Each technical chapter should begin by positioning its subject matter in the framework. The descriptions of PM characteristics and emission sources in nine regions should be presented by inserting regionally specific information into the same basic figure.

- The assessment authors do not appear to have made a clear and consistent distinction between public policy itself and policy-relevant science, and sometimes they unnecessarily shy away from interpretive statements, perhaps out of concern that they could be misconstrued as recommending policy. The committee recommends that each technical chapter (Chapters 2-9) conclude with a synthesis of how the scientific information presented could be used in managing ambient PM and what additional scientific information would be useful for making policy decisions in the future.

- The draft assessment focused much less on PM in Mexico than on that in Canada and the United States. For example, only one Mexican location, the Mexico City area, was included in the regional descriptions. The committee recommends that the assessment authors expand the representation of Mexico in the assessment, particularly by bringing in several recently completed and current studies in key Mexican cities, to fulfill NARSTO's goal of developing an inclusive North American perspective on PM.

- The assessment authors clearly recognize the importance of health effects associated with airborne PM, but they have failed to write a clear and readable summary of current PM-related health findings in the chapter devoted to the topic. The committee believes that a succinct summary of health effects would be useful for atmospheric scientists and recommends that the chapter be redirected to provide a framework for interactions between health scientists and atmospheric scientists and rewritten to reflect the numerous comments provided by the committee in Attachment B.

- Much attention is paid to the organic component of PM throughout the draft assessment, accurately reflecting recent research advances and the substantial remaining uncertainty on this topic. However, the committee found that the introductory material of the assessment does not call sufficient attention to this key foundation point. The committee recommends that a focused discussion of the importance of the organic component of PM be provided early in the assessment.

- The assessment authors gave inadequate attention to editorial details and the presentation of the text, particularly with respect to unnecessary repetition, inconsistent use of technical terms and concepts, and the quality of figures and tables. The committee recommends that there be a thorough editing to address those problems, as itemized in Attachment B. The authors should also strongly consider using a terminology box near the beginning of each chapter to define critical terms and abbreviations, which should also be consistently defined in a complete glossary and when introduced in the text.

COMMENTS ON THE EXECUTIVE SUMMARY OF THE DRAFT NARSTO ASSESSMENT

The draft NARSTO assessment begins with a long (53-page) executive summary intended for senior decision-makers and their advisers and the science community. As a preliminary step in the assessment process, NARSTO conducted interviews with senior decision-makers in Canada, Mexico, and the United States to help to elucidate policy goals for PM management, gaps in science, how science fits into decision-making, and how to present scientific information. On the basis of those interviews, NARSTO chose to use eight policy questions as the primary mechanism by which the draft executive summary would communicate the state of PM atmospheric science to decision-makers. The committee commends NARSTO for taking the time to query decision-makers; this step is often omitted in assessments. However, the committee has strong reservations about how the policy questions were developed. Furthermore, the responses to many of the policy questions and more generally the executive summary fall short of effectively communicating to decision-makers a clear description of the current understanding of PM atmospheric science and its policy implications. Specific recommendations for improving the executive summary include the following:

- The executive summary does not conform to standard expectations of executive summaries. It is too long and too technical for the intended decision-maker audience. A separate, brief executive summary that discusses in a terse manner the important points presented in the document should be written.
- A longer summary that more fully addresses the scientific issues and responds in detail to the policy questions is valuable. The current executive summary could meet this objective if renamed and extensively revised. In the body of the present review, the committee offers a suggested outline for revising the assessment's summary. The revised summary should introduce the charge, goal and specific objectives, and assumptions set forward by NARSTO for the assessment and then provide a crosswalk between the objectives and the main body of the assessment. The revised summary should also present a framework for informing airborne-PM management (as discussed above) and the adopted or proposed air-quality standards for PM for the three countries, which together form the foundation of the assessment.
- The interview process and method for developing the policy questions is not clearly articulated in the report and do not conform to current social-science methods. The policy questions were apparently developed before the interviews with decision-makers, who were then asked to confirm that the questions were appropriate. Thus, the policy questions are not necessarily the highest-priority questions about PM identified by policy-makers themselves but are the policy-relevant questions about PM that the assessment authors thought were the most important. In addition, only five of the 45 interviews were with Mexican decision-makers. Despite these problems, the policy questions are a useful organizing framework for communicating to policy-makers. The committee recommends that the eight policy questions be retained, with minor rewording to improve clarity, and that the method by which they were developed, including all its limitations, be clearly explained.

- The responses to the eight policy questions need to be revised. The committee recommends that each response begins with a clear and succinct reply to the question and follow it with a description of important scientific knowledge that supports it. Each response should also identify additional research needed to answer the policy question and to make better air-quality management decisions.

RECOMMENDATIONS FOR FUTURE ASSESSMENTS

The committee recommends that NARSTO, in preparing future assessments, enhance its interaction with the policy community. The committee finds that the draft assessment falls short especially in terms of the method for eliciting information needs from decision-makers and communicating the policy implications of the science to them. The committee strongly recommends that social scientists with expertise in elicitation of information be engaged in the process of developing policy guidance for future assessments. In providing policy-relevant atmospheric-science information, NARSTO should strive to discuss tradeoffs, options, and priorities more explicitly. By articulating policy implications less ambiguously, NARSTO will be more effective in informing policy development.

The committee also recommends that NARSTO focus on better placing the atmospheric-science information in the context of impacts on health, visibility, ecosystems, and global climate and with regard to implications for economics and other social sciences. To do that effectively, NARSTO needs to engage the relevant communities more fully and foster interaction with them. For example, NARSTO should consider establishing or increasing the prominence of leadership positions in its organization for persons whose responsibilities include linking to the other communities via standing committees, workshops, briefings, and similar endeavors. Failure of NARSTO to strengthen its cross-discipline interactions will limit the value of chapters of a future assessment, devoted to other related fields, as is the case with the chapter on health effects in the current draft assessment. What is more serious, the relevance of atmospheric-science research to the larger air-quality management activity could be compromised.

1

INTRODUCTION

Airborne particulate matter (PM)[1] is a critical issue for air-quality management in North America. Increasing evidence that human exposure to PM causes adverse public-health effects—especially respiratory and cardiopulmonary effects—has prompted regulators in Canada, Mexico, and the United States to reassess the adequacy of their standards (Department of Environment 2000; EPA 2002). All three countries are implementing or contemplating the adoption of more-stringent regulations on PM_{10} and on $PM_{2.5}$. In that context, NARSTO[2] has developed an assessment of the state of science pertaining to air-quality management of ambient PM.

Chartered in 1995, NARSTO is a public-private partnership with members from government, utilities, industry, and academe in Canada, Mexico, and the United States. Its primary mission is to coordinate and enhance policy-relevant atmospheric-science research and assessment on tropospheric ozone and PM. NARSTO's focus is on four broad technical fields: atmospheric chemistry and modeling research; emission research; monitoring, measurement, and observation-based analytic research; and integrated analysis and assessment. Initially, its efforts were directed at tropospheric ozone issues, and it produced an assessment of tropospheric ozone (NARSTO 2000). In 1998, NARSTO expanded its charter statement to include PM and began work on producing a PM assessment.

The NARSTO PM assessment was prepared by a team led by three cochairs (Peter McMurry, Marjorie Shepherd, and James Vickery) and including over 30 lead and contributing authors from the three nations. The specific charge, goals, objectives, and assumptions for the assessment are shown in Box 1-1. The assessment is intended to be a concise, scientifically credible, comprehensive discussion of the atmospheric-science issues associated with PM

---

[1] Airborne particulate matter refers to a broad class of discrete solid particles and liquid droplets of varied chemical composition and size. PM_{10} is defined as the mass of PM collected by a sampler with a 50% size cutoff at 10 μm in aerodynamic diameter; $PM_{2.5}$ is the mass of PM collected by a sampler with a 50% size cutoff at 2.5 μm.

[2] Originally, NARSTO stood for the North American Research Strategy for Tropospheric Ozone. When its charter was expanded to include PM, the term NARSTO became simply a word signifying the partnership.

BOX 1-1 NARSTO Particulate Matter Assessment Charge to the Assessment and Analysis Team

The Charge

The NARSTO charter includes collaboration of public-private research on particulate matter in air (PM) and calls for scientifically credible assessments and guidance for air quality managers and policy-makers. Therefore the Executive Assembly and Steering Committee charges the Assessment and Analysis Team to prepare an assessment (PM Assessment) of the state of scientific understanding of the atmospheric aerosol as it relates directly to policy questions and program management associated with implementing any new PM standards. Science topics to be addressed include air quality measurement methods and data, emissions information, atmospheric processes, and air quality modeling as they guide strategy development and implementation to reduce health and visibility impacts of PM.

The PM Assessment needs to be concise, scientifically credible, reasonably brief but comprehensive in its discussion, focusing attention on the strengths and weaknesses in current science underpinnings of air quality management tools. Specifically, the PM Assessment must contain an orderly presentation of the elements of the PM problem that starts with a definition of the problem and then lays out the issues, as policy questions and corresponding science question, to be resolved. This will include a discussion of where scientific knowledge appears to be sufficient, where important uncertainties lie and where future research would assist PM management in North America. Within this framework authors will explain why various scientific aspects of the PM issue are important to the policy community, provide direction as to what additional information could contribute to regulatory and other government decisions and thereby contribute to the overall priority setting for research within NARSTO. The PM Assessment is to be suitable for audiences consisting of air quality policy and management decision-makers, science-policy analysts, research managers, the science community, and the public. The publication of this PM Assessment report is targeted for the end of 2002.

Be mindful of the need to organize all of the scientific information presented such that it addresses and informs the policy issues facing environmental managers and regulators to the extent possible.*

Priorities should be tied to the decision-making process, that is, how will the new knowledge help make better decisions to improve air quality.

Provide explicit consideration of how implications of the science, or recent advances in the science, could apply to new approaches to reducing PM concentrations.

Be mindful of the treatment of interconnections among air quality issues: the multi-pollutant atmosphere.

_____

*Adapted from: Review of the NARSTO Draft Report: An Assessment of Tropospheric Ozone Pollution - A North American Perspective by the Committee

The NARSTO PM Assessment should attempt to evaluate critically the reliability and applicability of the technical and scientific tools currently available to support decision making for PM management. Discussion should include implications of current monitoring approaches, air quality modeling, and its inputs such as emissions, meteorological factors, and aerosol processing. The assessment also should address the requirements perceived to be needed for substantial improvement of these tools that are within reach of scientific investigations in the next five to ten years. The judgement of the meaning or definition of subjective terms like reliability and substantial improvement are left to the discretion of the team.

Goals and Specific Objectives

The overall goal for the PM Assessment is to fully describe how current knowledge and future research can aid air quality policy and management decision making. To satisfy this goal several specific objectives are to met.

1. Gain an understanding from **decision-makers** of information needs and constraints, including economic, policy, and implementation boundaries.

(Continued)

> **BOX 1-1** Continued
>
> 2. Provide a comprehensive conceptual model of aerosol formation and particulate matter distribution for **science-policy analysts and air quality decision-makers.** The model is to accommodate changing knowledge about atmospheric processes, emissions sources, emissions control technology, exposure, and human health and environmental impacts. It is to address existing limits in information and forecast the implications of expected results from on going and future research.
> 3. Provide a plain language conceptual description of particulate matter air quality for the **public** which describes the relevance of the atmospheric science research with its recent progress and findings.
> 4. Recommend atmospheric science and related emissions research, with priorities tied to the decision-making process, to **research managers** developing a coordinated research strategy for PM.
> 5. Provide a framework for **atmospheric scientists** that relates their work to standards, implementation, and air quality management and to health, exposure, and environmental impact research for standard setting.
> 6. Provide a context for **researchers in related fields** to link their work to that of the atmospheric science community, supplying important information on the current state of knowledge of particle formation and distribution and offering opportunities for future research coordination.
>
> **Assumptions**
>
> This charge to the Assessment and Analysis Team and the objectives are predicated on a number of assumptions about the assessment process and surrounding events likely to take place over the course of the assessment.
>
> 1. The definition of the problem is reducing levels of PM. Regulatory agencies are expected to recommend both PM<2.5 μm (fine PM) and PM 2.5 μm-10 μm (coarse PM) standards over the next two years. This assessment will encompass a review of PM characteristics relevant to both size fractions.
> 2. The PM Assessment will contain contextual information on exposure, health, and environmental impacts. This information will com from the extensive and in-depth science reviews, without update, prepared by Canada and the U.S. during their air quality goals and standards setting process.
> 3. A summary of related environmental issues, including deposition, climate change, and air toxins, is required to provide context for a discussion of the science of atmospheric aerosol occurrence and exposure.
> 4. Addressing PM, ozone linkages and other copollutant linkages are important to understanding the PM issue.
> 5. It is appropriate fro the PM Assessment to address contextual aspects of emissions control technologies mainly by reference to existing publications and information.
> 6. It is necessary to achieve a balanced assessment picture for the PM issues facing the NARSTO member countries.
> 7. The PM assessment process should be fully open to public participation.
> 8. The PM Assessment should explore accountability approaches that directly relate observations to determining the effectiveness of air quality management plans.
> 9. Given the continuing advancement of information on PM, this assessment will be the first in a series of periodic NARSTO assessments.

management. As such, the draft assessment focuses on measurements, emission data, atmospheric processes, and air-quality models. Summaries of known and potential effects on human health and visibility are also included to provide context for concern about PM. Some analysis of the technical information is provided via recommendations for future research needs

INTRODUCTION 9

and "conceptual descriptions" of the factors that affect observed ambient PM concentrations in nine regions in North America. The executive summary of the assessment contains a summary of the information presented in the body of the document and responses to eight policy questions, which are used to synthesize the atmospheric-science information necessary for implementing ambient PM standards. The assessment was intended to be written from a policy-relevant perspective and to develop a common scientific understanding of assessment tools that can be used in implementing and attaining PM ambient air-quality standards.

The National Research Council (NRC), in consultation with the Royal Society of Canada (RSC) and the United States-Mexico Foundation for Science (Fundacion Mexico Estados Unidos para la Ciencia, FUMEC), was asked to review NARSTO's draft PM assessment, dated December 31, 2001 (referred to in the present report as the assessment or the draft assessment). In response to that request, the NRC established the trinational Committee to Review NARSTO's Scientific Assessment of Airborne Particulate Matter (referred to in the present report as the committee). Box 1-2 shows the charge provided to the committee. The committee was formed with input and concurrence from RSC and FUMEC in accordance with a memorandum of understanding that was approved by the two organizations and the NRC. The review was conducted in accordance with standard NRC policies, including compliance with Section 15 of the U.S. Federal Advisory Committee Act. Previously, the NRC reviewed a draft of NARSTO's tropospheric ozone assessment (NRC 2000).

In the present review, the committee provides comments on how well the NARSTO assessment meets its overall objective and goals and offers detailed technical review comments. The review begins with general comments on the assessment (Chapter 2), which is followed by a critique of its executive summary (Chapter 3), and then comments on individual chapters (Chapter 4). Recommendations for future NARSTO assessments are provided in Chapter 5. A list of line-by-line editorial comments is provided in Attachment B of this report. The committee reviewed the NARSTO assessment in light of its members' collective knowledge of other PM assessments and relevant reports (e.g., NRC 1998, 1999, 2001; EPA 2002; IPCC 2002; Molina and Molina 2002). The committee did not attempt to make specific comparisons with those other documents.

BOX 1-2 Statement of Task for Committee to Review NARSTO's Scientific Assessment of Airborne Particulate Matter

An NRC committee will review a draft version of NARSTO's Fine Particle Assessment in cooperation with the Royal Society of Canada and the United States-Mexico Foundation for Science. The NRC will conduct the review in accordance with standard NRC policies, including compliance with Section 15 of the Federal Advisory Committee Act as it applies to the U.S. National Academy of Sciences. As part of its review, the committee will take into consideration the stated overall goals and objectives of the assessment document. The committee will also consider this assessment in light of other particulate matter assessments that have been completed in the past few years. In addition, the committee will take into consideration the reports of relevant NRC/NAS committees, including the Committee on Research Priorities for Airborne Particulate Matter, as well as the work of other organizations.

As part of its review of the NARSTO assessment document, the review committee will include consideration of the following criteria:

(Continued)

> **BOX 1-2** Continued
>
> Are the goals and objectives of the assessment clearly described and fully addressed in the document? Does the assessment go beyond its scope?
>
> Do evidence and analysis adequately support the conclusions and recommendations? Are uncertainties or incompleteness in the evidence explicitly recognized? If any recommendations are based on value judgments or the collective opinions of the authors, is this acknowledged and are adequate reasons given for reaching those judgments?
>
> Are the data and analyses handled adequately? Are statistical methods applied appropriately?
>
> Are policy-relevant questions appropriately addressed? Are the advantages and disadvantages of alternative options, including the status quo, considered?
>
> Are the presentation and organization of the document effective?
>
> Is the document impartial?
>
> Does the executive summary concisely and accurately describe the key findings and recommendations? Is it consistent with other sections of the report?
>
> What other significant improvements, if any, might be made in the report?
>
> In providing comments, the review committee is encouraged to distinguish issues it considers to be of major concern from other, less significant points.

2

OVERARCHING ISSUES

The committee finds that the draft assessment is a good draft with respect to the atmospheric-science topics discussed and the level of detail. However, it needs improvements to meet its goal of communicating well with decision-makers. Of greatest importance is the committee's recommendation that a framework for understanding the complexities of atmospheric PM and its management be more clearly and consistently presented to provide the reader with a context for the assessment's subject matter. The assessment also needs substantial improvements in presentation. It is long, repetitive, and in some places inconsistent. Many figures and tables are in poor condition. In the following sections, the committee provides overarching comments directed at the assessment's contents and presentation.

PM CONCEPTUAL MODEL AND FRAMEWORK FOR INFORMING AIRBORNE-PM MANAGEMENT

The idea of a conceptual model, typically a simplified and general description of complex processes, is used in a number of ways throughout the PM assessment. The loose definition and multiple applications of the term create confusion for the reader. For example, Figure 10.1[1], "Elements of a conceptual model for PM adopted in this study," is presented in Chapter 10 and in the response to policy question 3. The figure provides a highly simplified schematic of the factors that affect PM. Figure 6.8 in Chapter 6 shows another type of conceptual model, which is said to be useful for evaluating the effect of reducing sulfur in Canadian gasoline; it provides specific information linking sources and ambient concentrations. A third form of conceptual model is defined and presented in Chapter 10 for each of the regional case studies; the idea of a "conceptual description," which is defined as an incomplete conceptual model, is introduced here. The distinction seems unnecessary, given that conceptual models are typically not expected

---

[1] References to figures and tables are to those in the draft NARSTO PM assessment unless specified as being "in the present report" or the like.

to be complete. Other definitions and examples of conceptual models are provided in Sections 1.8 and 9.2.3.

Given the multiple uses of "conceptual model" in the assessment, the committee struggled to determine how NARSTO intended to use the term. Looking to the goals and specific objectives that NARSTO set out for the assessment (see Box 1-1 in Chapter 1 of the present report), the committee notes that the second objective is to "Provide a comprehensive conceptual model of aerosol formation and particulate matter distribution for science-policy analysts and air quality decision makers." However, although conceptual models may be comprehensive in the categories they include, they are by definition not comprehensive in detail, so this goal would be rather difficult to reach. A single conceptual model explicitly linking sources, chemical and physical processes in the atmosphere, and the environmental distribution of PM would be useful, but it is not identified in the assessment. What is provided in the assessment is a set of regionally specific "conceptual descriptions" that are neither comprehensive in the regions evaluated nor complete in the processes described. A single general conceptual model for airborne PM would go a long way toward meeting NARSTO's second objective as well as its third objective, to provide a plain-language conceptual description of PM air quality for the public.

The committee recommends that a single general conceptual model for airborne-PM burden be explicitly introduced in the beginning of the assessment. The conceptual model could readily include emissions, atmospheric processing, and loss processes for airborne PM, all of which can be influenced by meteorologic factors and ultimately affect the distribution of PM in the atmosphere. To alleviate confusion, "conceptual model" should be clearly defined and consistently used throughout the document, and "conceptual descriptions," "conceptual understanding," and similar uses of "conceptual" should be eliminated. It should be made clear in the introduction of the conceptual model that it includes generalized concepts of major factors involved in the generation and distribution of PM.

NARSTO's second objective also calls for the model to "accommodate changing knowledge about atmospheric processes, emission sources, emission control technology, exposure, and human health and environmental impacts." Similarly, the fifth objective calls for the assessment to "provide a framework for atmospheric scientists which relates their work to standards, implementation and air quality management, and to health, exposure, and environmental impact research for standard setting." Thus, effectively connecting the conceptual model of airborne-PM burden to the impacts and policy implications of increased PM requires placing the conceptual model in a larger framework for informing airborne-PM management. Figure 1.6 in the draft assessment illustrates part of the interaction between science and policy, but it is not detailed enough to meet the objectives.

The committee recommends that a clear framework for informing airborne-PM management, of which the conceptual model for airborne-PM burden is an important component, be explicitly introduced in Chapter 1 and near the beginning of the executive summary. The framework should be sufficiently fundamental for its components and their implications to be readily understood by lay readers (that is, it should be largely self-explanatory) and to serve as a basic paradigm for understanding variations that apply to any region over time. An example of such a framework is provided in Figure 2-1 of the present report, and the committee recommends that this framework or something similar of NARSTO's choosing be used in the assessment. The framework in Figure 2-1 of this report includes on the left side (first three columns) a single general conceptual model for airborne-PM burden; the right side includes a fourth column devoted to assessment activities that seek to understand the sources of observed PM concentrations, the impacts of PM, and the expected outcomes of various management strategies. The information gathered in the assessment column feeds into the management of PM, as shown in the fifth column. The committee recognizes that this proposed framework extends beyond the scope that NARSTO defined for this assessment—for example, the framework includes impacts on acid deposition, climate, and ecosystems—and is not suggesting that discussion of these topics be added to the current assessment. Nonetheless, providing a broader context by using a more complete framework would be useful for the intended audience of the assessment.

In introducing the framework and discussing it throughout the assessment, the authors should emphasize that the framework, especially the conceptual model for airborne-PM burden, can be adapted for specific cases, such as those in Chapter 10 of the assessment, by obtaining appropriate information for each of the components. Variations of specific factors over time and space will cause the different components to be more or less important for a given case. Moreover, although example cases can be presented, the actual combinations of factors at play allow for essentially infinite variations.

The committee views the establishment of a single general framework for informing airborne-PM management as critical to the success of the assessment, which depends on how readily the reader can assimilate the detailed information presented throughout the document into a single, consistent conceptual structure. Establishing a sound framework early in the document allows sections to refer to it to show the relevance of expanded information. The committee recommends that each technical chapter begin by positioning its subject matter in the framework for airborne-PM management. Likewise, the information on regionally specific processes could be presented using the same basic figure as the framework but with information in each box reflecting regional characteristics. Such a figure for each region that brings together all the factors affecting its PM problem may be more effective than the tables now presented in Chapter 10, where the information is grouped by factor (such as sources, meteorology, and policy implications).

OTHER CONTENT ISSUES

One of the stated charges to the team that prepared the draft assessment was to assess the "state of scientific understanding of the atmospheric aerosol as it relates directly to policy questions and program management associated with implementing any new PM standards." However, the authors do not appear to have made a clear and consistent distinction between public policy and policy-relevant science, and sometimes they unnecessarily shy away from interpretive statements, perhaps out of concern that they may be misconstrued as policy recommendations. One way to characterize this crucial distinction is that public policy is the articulation of the hierarchy of preferred decision options, the range of instruments that can be used to implement options, and the rationales for choosing among the options, whereas policy-relevant science informs or constrains the nature of the options. The committee recognizes that NARSTO's objective is not to promote specific policies but to summarize the policy-relevant

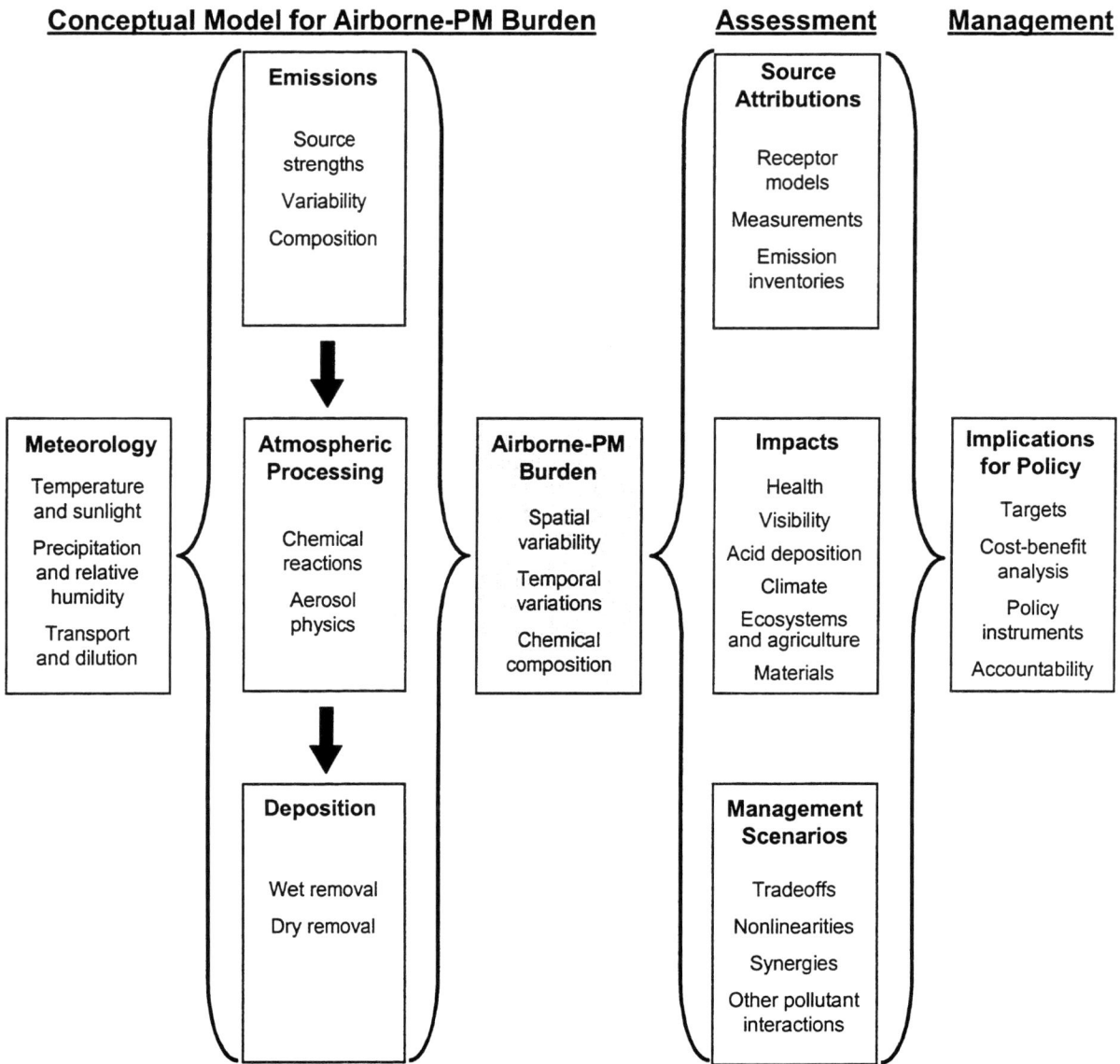

FIGURE 2-1 Framework for informing airborne-PM management.

science. The assessment, however, often lacks a clear discussion of how the material presented could be useful for informing PM policy decisions. The committee recommends that each technical chapter (Chapters 2-9) conclude with a synthesis of how the scientific information presented could be used in managing PM. Such a "policy implications" section should follow the technical summary at the end of each chapter.

Another issue is that the assessment focuses much less on airborne PM in Mexico than on that in Canada and the United States. The assessment includes only one Mexican location—the Mexico City area—in its regional descriptions, although information on PM in other areas of

Mexico is available. Likewise, only five of 45 interviews in the development of the eight policy questions were with Mexican decision-makers; this potentially compromises the usefulness of the questions and responses to Mexican policy-makers. Expanding the participation and representation of Mexico in the assessment would improve its applicability to PM management there and help to fulfill NARSTO's goal of developing a North American perspective on the PM problem. Two NARSTO special journal issues have relevant studies in Mexico that can be referenced (Watson et al. 2001; Chow et al. 2002). In addition, several recently completed or current studies focus on key Mexican-U.S. border regions, and the data from and implications of these studies would be appropriate and pertinent; the studies include Mejia-Velazquez and Rodriguez-Gallegos (1997), Mukerjee (2001), Molina and Molina (2002), and those listed on the Website of the Southwest Center for Environmental Research and Policy (SCERP 2002).

The need to understand the organic component of PM is a general theme that runs through the document. The committee agrees that a better understanding of this component of PM is appropriate, given that it is a large portion of ambient PM and that major uncertainties are associated with it. However, the authors need to motivate this focus better from the beginning of the report, perhaps through the use of a text box in Chapter 1, by summarizing why improving the characterization of organic carbon PM is critical for managing PM. In particular, the challenges posed by the large number of organic species and the present inability to resolve organic matter content completely should be explained.

Another issue that needs more careful attention in the assessment is spatial variability in $PM_{2.5}$ composition and concentrations. Epidemiologists have increasing evidence that health effects of $PM_{2.5}$ and its components vary over spatial scales smaller than that resolved by current monitoring networks. The committee recognizes that the available monitoring network limits the extent to which smaller-scale variations in $PM_{2.5}$ can be presented in the assessment. However, the draft could leave the impression that $PM_{2.5}$ concentrations and composition tend to be rather homogeneous across regions in the size range of those highlighted in Chapter 10. Furthermore, the potential need for more fine-scale $PM_{2.5}$ measurements is not noted. The committee recommends that discussion in the assessment reflect the heterogeneity in the composition and concentrations of $PM_{2.5}$ within a region, the potential effect of this heterogeneity on variability in exposure and health effects, and the possible need to obtain additional measurements to support health-effects research.

Finally, the extent to which the draft assessment represents the views of the North American atmospheric-science community is not clear. It is uncertain what sort of guidance was provided to chapter authors and whether reviewers with particular technical expertise were recruited to conduct scientific reviews of the chapters. In addition, it is unclear to what extent the research recommendations provided in Chapter 11 represent the views of the NARSTO community as a whole. Given the importance of those recommendations for defining research objectives, it is important for the authors to discuss how they were developed.

PRESENTATION ISSUES

The committee recommends that the assessment document be thoroughly edited to reduce repetition and ensure that terms and concepts are used consistently throughout. Numerous examples of those problems are presented in this review's comments on the draft's chapters. In addition, although some chapters are well written, others would benefit greatly from copyediting to improve distracting grammatical and usage errors. Besides a thorough editing, the authors should also strongly consider using a text box near the beginning of each chapter to introduce and define important terms, abbreviations, and concepts. Such boxes, which are more immediately accessible than a glossary, should supplement definitions given in the glossary. In addition, consistent definitions should be provided in the text as terms are introduced. It is important that the definitions used at different points be the same and that the glossary be complete.

The committee finds the presentation of major research recommendations in text boxes confusing. There is no discussion in the draft's introduction alerting the reader that important research recommendations are summarized in this manner and that the numbering scheme used for the text boxes refers to the numbering used in Chapter 11. The specific text of the recommendations given in the chapters often does not match the text in Chapter 11. Many of the cross references to report sections used in Chapter 11 to support the recommendations are to nonexistent or irrelevant sections. The committee recommends that these problems be corrected. Using text boxes to highlight only research recommendations may give the reader the impression that the authors give more weight to these recommendations than to summarizing policy-relevant PM science. In fact, the authors could consider the potential value of using such boxes to highlight the policy recommendations presented in Chapter 10.

In general, the figures and tables are in poor condition. Many have inadequate captions or titles, missing labels, and poor reproduction. In addition, because they typically have been excerpted from other published materials, they are often very technical and inadequately explained in the text. The committee recommends that each figure and table be examined to ensure that its caption or title is adequate, that all text and lines are visible, and that it is sufficiently supported in the text of the assessment. Specific figures and tables that are particularly poor in presentation or are not well integrated into the discussion are noted in Chapters 3 and 4 of this report.

A consistent level of citations to the relevant technical literature should be adopted. The committee agrees that the PM assessment should not be an exhaustive review of prior work, but several chapters would benefit from additional references to key publications and reviews. Overall, the committee finds that inadequate attention has been given to editorial details and the presentation of the draft report. It was disappointing to note that many of the committee's specific editorial comments had already been conveyed by chapter authors and members of the NARSTO community but were not addressed before the draft was provided to the committee.

3

COMMENTS ON THE ASSESSMENT'S EXECUTIVE SUMMARY

The NARSTO PM assessment draft begins with a 53-page "Executive Summary." The assessment cochairs indicated to the committee that the summary would be published with the full document and also separately in English, Spanish, and French (Hales 2002; Shepherd 2002; Vickery 2002). The executive summary is currently structured with several distinct sections:

- A short introduction.
- A list of 19 "key insights" considered to be the most important scientific results that decision-makers should know.
- Answers to eight policy questions (PQs) with some figures and tables drawn from the chapters and indications of where in the chapters the supporting evidence could be found.
- Five major recommendations, which pertain primarily to research needs.
- A narrative titled "Highlights of the Assessment."

The assessment cochairs indicated that they chose to structure the executive summary in this way so that it would be capable of speaking to three audiences: senior decision-makers, scientific advisers to the senior decision-makers, and the science and science-user community (M. Shepherd, Meteorological Service of Canada, personal communication, April 26, 2002).

RESTRUCTURING OF THE EXECUTIVE SUMMARY

The committee finds that the executive summary as written does not conform to standard expectations of executive summaries in that it is too long and too technical for the intended decision-maker audience. A separate, very brief executive summary that discusses in a terse manner the important points presented in the document should be written. Nevertheless, a longer section that is similar to the current executive summary, that treats the issues more fully, and that responds in more detail to the PQs is valuable, particularly if it is aimed at the scientific advisers to decision-makers. This longer section could be called "Synthesis of Key Issues" or have some other title of NARSTO's choice. Both the brief executive summary and the longer synthesis need to be written in plain language, avoiding atmospheric-science jargon as much as possible. In the synthesis of key issues, the objectives of the assessment and its limitations and assumptions need to be clearly described. The committee recommends that the charge, goals and specific objectives, and assumptions set forth by NARSTO for the assessment (see Box 1-1) be included in their entirety in the synthesis. At present, that information appears nowhere in the document. Without such material, the reader is uncertain why this assessment was written and what its intended audience is.

Of particular importance is an explicit definition of the "PM problem" in the context of NARSTO's first assumption for the assessment (see Box 3-1 in the present report). At no point in the current executive summary is it clearly and explicitly stated how the PM problem is defined. When the committee queried the assessment cochairs, they indicated that they considered the PM problem to be exceedance of existing or proposed mass-based standards. The committee appreciates why the authors focused on providing scientific and technical guidance toward meeting the standards, but this substantial limitation of the scope needs to be made clear.

In addition to clearly stating the objectives of the assessment, the synthesis of key issues should provide a crosswalk between the objectives and the main body of the assessment. One way to accomplish that would be to align the discussion in the synthesis with the objectives. The committee has developed an outline that shows how this could be done (see Box 3-1). According to the suggested outline, the synthesis of key issues would begin with a discussion of why the assessment was written and a presentation of the charge, goal, objectives, and assumptions that guided and limited the preparation of the document. The second section of the synthesis would explain the PM problem in terms of a framework for informing airborne-PM management, thereby addressing NARSTO objectives 2 and 3. As discussed in more detail previously (in Chapter 2 of this report), the framework should enable a reader to understand what PM is, how the various processes that influence it interact, and what is involved in assessing and managing the PM problem. The third section would describe how the task was approached, with a focus on explaining the interactions with the policy community, which is NARSTO objective 1. The fourth section would step through the eight PQs, providing answers that would be useful to research managers (NARSTO objective 4) and linking the answers to specific atmospheric-science research accomplishments and needs (NARSTO objective 5). The last section of the synthesis would address NARSTO objective 6 by providing a context for researchers in related fields.

OTHER COMMENTS ON THE EXECUTIVE SUMMARY

Elicitation of the Policy Questions

The committee commends NARSTO for taking the time to query decision-makers about their scientific and technical needs. Efforts to identify what information would be useful to the intended audience before writing a scientific assessment of this sort are rare. A document providing the information that decision-makers expressly request is certainly valuable, and presenting this information as responses to specific PQs is a useful format.

BOX 3-1 Suggested outline for a revised NARSTO PM Assessment

Preface: What NARSTO is and why it is an appropriate group to do an assessment of airborne PM

Executive Summary: A few pages of the most important messages for policy-makers

Synthesis of Key Issues:

1. Discussion of why the report was written, highlighting the need to assist the policy community in interpreting complex and new scientific information associated with airborne PM
   a. Motivation
   b. Charge *(see Box 1-1 of this report)*
   c. Goals and the six specific objectives *(see Box 1-1 of this report)*
   d. Assumptions *(see Box 1-1 of this report)*
   e. Framing of the problem in the context of the air-quality standards (introduced here)

2. Explanation of the PM problem via a framework for informing airborne-PM management accompanied by plain-language discussion and multiple figures *(NARSTO objectives 2 and 3)*
   a. Rationale of the framework
   b. Factors that affect PM, including anthropogenic and natural sources, seasonality, and regional and local contributions (use existing text from the response to PQ3)

3. Description of the exercise and of how the task was approached
   a. Interaction with policy community *(NARSTO objective 1)*
      i. Approach to eliciting input from decision-makers
      ii. Scope of decision-makers polled
   b. Approach to reviewing scientific information, selecting authors, and choosing citations
   c. Outline of the full assessment document

4. Review of issues framed by the PQs, addressing all the following topics for each question *(NARSTO objectives 4 and 5)*
   a. Current answer to each question, stated clearly and succinctly
   b. Key knowledge that supports the answer, highlighting recent findings and advances (use existing text from "Key Insights" and "Highlights of the Assessment")
   c. Additional atmospheric-science knowledge needed to answer the question better (use existing text from "Recommendations")
   d. How the recommended research may affect air-quality management decisions

5. In the overall context of the eight PQs, discussion of how new atmospheric-science knowledge would inform research in other fields (including health, ecosystems, climate, acid deposition, and visibility) and how research in those fields can help to guide atmospheric-science research *(NARSTO objective 6)*

Body of the Report

Appendixes

The committee is concerned, however, that the method for selecting and formulating the PQs is not clearly articulated and thus lacks transparency in the assessment document. To assist in the committee's deliberations, the assessment cochairs provided a summary of how the eight PQs were identified (M. Shepherd, Meteorological Service of Canada, personal communication, April 26, 2002). The committee's understanding of the process is that the assessment cochairs distilled the PQs from a larger set of questions developed at a number of NARSTO workshops. The set of eight PQs was refined by the assessment cochairs with input from NARSTO members and finalized at the first meeting of the assessment authors.

Later, the assessment cochairs conducted interviews with senior decision-makers in federal, state, and provincial environment departments and in private industry in Canada, Mexico, and the United States. The interviews included questions about policy goals for PM management, gaps in science, how science fits into decision-making, and how to present scientific information. As part of the interview, decision-makers were asked to comment on whether the eight PQs "capture the issues that need to be resolved." On the basis of the comments, the assessment cochairs decided that the decision-makers had confirmed the usefulness of the eight PQs previously developed.

The committee finds the method used to select and formulate the eight PQs unsatisfactory. The approach does not conform to current social-science methods. The potential stakeholders assembled for interviews were not representative. Only government and industry representatives were selected, and only five of the 45 interviews were with decision-makers from Mexico. The resulting understanding of what information decision-makers need from the atmospheric-science community may be inadequate.

Another deficiency is the apparent lack of a well-constructed interview protocol. Several of the questions posed to the decision-makers called for a mix of prompted and unprompted responses. For example, one question asked of decision-makers was "What issues need to be resolved to achieve these [policy] goals?" In providing some context for the question, the interviewers listed the eight PQs and asked whether they "capture the issues that need to be resolved." A respondent would probably be confused about how to answer the initial question, particularly about whether to contribute a separate list of issues or to simply affirm that the list provided was sufficient. As a result of that deficiency, the eight questions do not necessarily represent the highest-priority questions about PM identified by policy-makers themselves but, rather the policy-relevant questions about PM that the assessment authors thought were the most important, which could be substantially differentLikewise, as far as the committee can tell, no rigorous quantitative analysis was used to assess the responses. In their presentation to the committee, the assessment cochairs showed only bar graphs tabulating responses by decision-makers to various questions (Vickery 2002). The total number of responses varied greatly for different questions. Indeed, it appears difficult to tabulate answers to questions like the example above because some respondents might contribute their own list of issues and others might just affirm that the provided list was sufficient. If a rigorous quantitative analysis was conducted, the committee recommends that the method be transparent in the assessment document and that the results be presented in the synthesis of key issues.

Despite the problems with how the PQs were obtained and characterized, the committee recognizes that the exercise of generating them cannot be redone at this point. Furthermore, despite their limitations, the PQs appear generally appropriate, with some exceptions that are noted below. The committee recommends that the PQs be retained as long as the method by which they were developed, including all its limitations, is clearly explained (Part 3a in the revised outline). It is particularly important to state that the PQs were developed by the assessment team, which later sought confirmation from decision-makers about their policy relevance. The committee strongly recommends that social scientists with expertise in elicitation of information be engaged in the process of developing policy guidance for future assessments conducted by NARSTO.

Introduction of the PM Standards

Despite defining the PM problem in terms of exceeding existing or expected standards for ambient PM concentrations, the executive summary does not clearly introduce the standards in the three nations. The discussion in response to PQ1 does mention where observed concentrations of PM exceed the national standards; however, this discussion is confusing because there are many different standards. For example, the U.S. 24-h $PM_{2.5}$ standard of 65 $\mu g/m^3$ is higher than the Canadian 24-h $PM_{2.5}$ standard of 30 $\mu g/m^3$, and Mexico does not have a standard for $PM_{2.5}$.

The committee recommends that the standards, when they became or will become effective, and whether or when they will be reviewed be presented in the synthesis of key issues. The discussion should clarify that different sizes of PM are regulated, that the standards have different averaging times and statistical forms, and that each nation has set standards to address multiple regulatory goals (such as, protecting human health and improving visibility). A table comparing the different standards may be effective for summarizing the information. It should also be mentioned that although the PM problem is defined in the draft assessment in the context of the standards, health problems associated with PM have been detected at concentrations below those of the standards (Samet et al. 2000). In addition, health problems associated with PM likely exist in countries where standards have not been set, and the document should not be written in a way that may imply the contrary.

Integration of "Key Insights" and "Highlights of the Assessment" into Discussion of Policy Questions

The executive summary starts with a list of 19 "key insights" that are intended to provide the "most important scientific aspects of current knowledge relevant to the atmospheric-science underpinnings of multinational PM-management strategies" (M. Shepherd, Meteorological Service of Canada, personal communication, April 26, 2002). Although the key insights are interesting and useful, the committee finds that having them as the first section of the executive summary is not effective for communicating them to decision-makers, who may not at that point in their reading of the document have a basic conceptual understanding of PM. The insights should be presented in such a way as to clearly indicate what scientists do and do not know, to be connected to the PQs and to the framework for airborne-PM management, and to provide a rationale for research recommendations. The final section of the current executive summary is a narrative titled "Highlights of the Assessment." The committee finds that this section is generally well written but that it contains much of the information presented elsewhere in the executive summary. Some of the text in the highlights section could be used in section 2 of the synthesis of key issues, where the framework for informing airborne-PM management is described. As in the case of the key insights, the ties between the information presented in the highlights section and the PQs, the general framework for informing airborne-PM management, and the research recommendations could be clarified.

To address those concerns, the committee recommends that the key insights and some of the material in the highlights section be incorporated into the responses to the eight PQs, as indicated in the suggested outline of the synthesis of key issues presented in Box 3-1. The response to each PQ needs to include the scientific knowledge available to answer the question, which is basically the information provided in the current key insights. Presenting this scientific information with the PQ will clearly demonstrate that recent scientific advances relevant to policy needs have been made. Describing the scientific knowledge that is available to answer the PQs could easily lead into a discussion of gaps in the understanding of PM and the need for additional research. Indeed, the recommendations listed on the current pages ES-32 to ES-33 could be presented in text boxes accompanying the discussion, thereby highlighting the connection between the PQs and the research needed to answer them.

Use of Figures and Tables

A number of figures and tables are presented in the executive summary to illustrate points made in the text. Most of the figures are brought forward from the body of the draft assessment, where they are discussed in more detail, and many are reproductions of figures used in other publications. Although the figures and tables provide useful additional information and enhance communication of many issues, the committee finds that they are not adequately integrated into the executive summary. The reader is left to determine the relevance of most of them and to identify the important information they convey. For the more technical figures, it is also important to provide more complete explanations in the captions, as was done well for Figure 6.4 (used to address PQ4).

COMMENTS ON THE POLICY QUESTIONS

The eight PQs are the primary mechanism by which the draft executive summary communicates to decision-makers. The committee directed substantial energy to assessing the questions and the responses to them. As discussed in the previous section, the committee has some reservations about how the PQs were developed and used, but it finds that they are a useful set of questions nonetheless. Some suggestions for rewording the questions are given below, but the committee recommends that the general intent of each question be retained. The committee recognizes that rewording the questions presents some difficulty for NARSTO in explaining its prior elicitation activities, but feels that the improvement in communication gained by clarifying the questions outweighs this concern. In general, the committee finds that presenting the policy-relevant information in a question-answer format is effective for communicating to decision-makers.

Many of the responses to the questions should be revised. In general, each response should address the four topics described in Part 4 of the suggested outline for the synthesis of key issues (see Box 3-1). A response should begin with a clear and succinct answer to the question and follow it with a description of important scientific knowledge that supports that answer. The explanation of scientific knowledge should highlight recent findings and advances and identify gaps in knowledge that call for additional research. Finally, the response should explain how additional recommended research would help to answer the PQ and would affect air-quality management decisions.

Another general concern about the presentation of the PQs pertains to the subquestions listed immediately below most of the questions. The committee is uncertain about the intent of the subquestions. They appear to be intended to clarify the question or define the scope of the response. But the subquestions are not always addressed, as in the case of PQ3. The committee recommends that the intent of the subquestions be made clear if they are to be retained. Other options are to use the ideas expressed in the subquestions to improve the questions or to remove the subquestions.

The committee also cautions the authors to be careful in their use of first-person pronouns in the PQ discussion. The current text uses "we" in the phrasing of the questions and in some of the responses. Employing the first-person pronoun contributes to the confusion because it is unclear whether it refers to NARSTO, the research community more generally, environmental regulatory agencies, decision-makers, all these, or some other entity.

A final concern about the overall presentation of the PQs involves how PQ2, PQ3, and PQ4 fit together. Those three questions characterize the PM problem, its sources, and possible options for addressing it. The current responses do not align exactly with the questions, indicating that perhaps the intent of the questions is not clear. The committee can conceive of at least two ways to interpret how the three questions are supposed to interconnect. PQ2 could be intended to describe the composition of PM at various locations, PQ3 to link sources to the composition of PM to its likely sources, and PQ4 to identify possible control options for the sources. Or PQ2 could be intended to describe the composition of PM and its likely sources, PQ3 to discuss broad control strategies that could be implemented nationally or over large regions, and PQ4 to identify sources that need to be controlled locally. The committee recommends that the three questions be clarified and that the responses given address them directly.

PQ1: "Do we have a significant PM problem, and how confident are we?"

This question is framed correctly and is policy-relevant. Assuming that the "PM problem" is more clearly defined earlier in the document, the question should not need to be rephrased except to clarify to whom "we" refers.

The current response to PQ1 does not provide a clear answer. The committee recommends that the answer be "yes; very confident" for the following reasons:

- Persistent exceedances of the existing and proposed standards have been observed.

Some information about observed ambient PM concentrations in all three nations should be presented here so that the reader can compare these values to the standards.

- Large portions of the population in North America are exposed to PM concentrations above the current or proposed standards. To emphasize the point and to provide context, it would be useful to include statements such as, "There are X million people in Canada, X million people in Mexico, and X million people in the United States that live with average PM levels above Y $\mu g/m^3$."
- There is considerable evidence suggesting health effects are associated with exposure to PM. A brief summary of the health effects should be provided, including discussion of some of the subtleties associated with the standards, such as these:
  - Thresholds for population health effects have not been observed (Schwartz et al. 2001; Dominici et al. 2002).
  - There may be regional differences in the potency of PM (Samet et al. 2000) and these may be associated with compositional differences.
  - There is evidence of health effects of acute, short-term exposure and long-term exposure to PM (Dockery et al. 1993; Pope et al. 1995; Abbey et al. 1999; Samet et al. 2000; Pope et al. 2002).
  - There are a few examples of relationships between reduced PM exposure and improved health (Pope 1989; Chay and Greenstone 1999; Friedman et al. 2001; Avol et al. 2001).
- In addition to health effects, high concentrations of ambient PM are associated with acid deposition and the resulting deterioration of ecosystems and structures, visibility impairment, and effects of climate change. All those impose societal costs, so controlling PM will yield economic and other benefits beyond improved health. Of course, it should be pointed out that the assessment focuses on human health effects and visibility impairment.

A better discussion of the standards in the synthesis of key issues will clarify the response to PQ1. Because the PM problem is defined for the assessment as exceedance of the existing or proposed standards, it is important that the standards be clearly presented to answer this question. The current response does not provide any information about ambient $PM_{2.5}$ concentrations in Mexico. Some data are available, including those discussed in Chapter 10 of the draft and additional studies (e.g., Mejia-Velazquez and Rodriguez-Gallegos 1997; Mukerjee 2001; Molina and Molina 2002; SCERP 2002).

PQ2: "Where we have a PM problem, what is the source of the concentrations we observe?"

This question is important for policy and could be expressed more precisely. The current response is more consistent with the question "Where we have identified a PM problem, what is the composition and what are the sources of the elevated concentrations that are observed?" A general response to this rephrased question is given in the second bullet, which qualitatively describes the composition and broad source categories for various locations. That discussion is accompanied by two figures, one that shows the composition of $PM_{2.5}$ mass at 16 locations and a second that shows the contribution of local and regional sources of $PM_{2.5}$ at seven locations.

Overall, the committee finds that the discussion and figures are a good response to the rephrased question.

One problem with the response is that it is not sufficiently specific about the types of sources. For example, it is stated that 95% of SO_2 emissions, the source of particulate sulfate, comes from fossil-fuel combustion, but the proportional contribution of electricity generation to such emissions is not described. From the perspective of PM management, more information about the sources of specific emission sources could help decision-makers to target the control strategies more effectively. A second problem pertains to the lack of discussion about Mexico, aside from the inclusion of one city in Figure 5.12. The assessment should make it clearer where a PM problem has been identified, but observational data are insufficient to determine the source of the problem. Given the high concentration of PM reported in Mexico, the committee finds that more emphasis on cities other than Mexico City, especially the border region between Mexico and the United States, is warranted (e.g., Mukerjee 2001; Mejia-Velazquez and Rodriguez-Gallegos 1997).

PQ3: "What broad approaches might we take to fix the problem?"

This question is appropriate and policy-relevant, but it could be rephrased to clarify the distinction between the "broad approaches" discussed here and the "specific options" discussed in the response to PQ4. Furthermore, the committee notes an inconsistency between how PQ3 is phrased in the draft assessment and in the presentation given by the cochairs at the first meeting with the committee (Vickery 2002), in which the question was phrased as "What broad *(pollutant based)* approaches might we take to fix the problem?" (italics added).

The confusion over the definition of a "broad approach" is amplified by the lack of a definitive answer to the question. The first bullet of the response discusses generally how one might develop a strategy to control PM. The second bullet provides a partial description of the conceptual model for airborne-PM burden coupled with some tentative suggestions about emission-reduction strategies; much of this text belongs in Section 2b in the synthesis of key issues, where the conceptual model is described. The last bullet gives specific recommendations for the individual case-study areas and should instead be used to answer PQ4. The sub-bullets have too much detail to be included as broad approaches.

The committee recommends that the response be rewritten to elaborate that a large portion of the $PM_{2.5}$ mass loading is anthropogenic. In particular, the response should identify the major anthropogenic sources. Figure 6.7 and the table from slide 12 of Shepherd (2002) may be appropriate for this purpose. In addition, the committee recommends that the response try to identify source categories that might be good targets for broad (nationwide or at least regional) emission control.

The committee also finds that it may be appropriate to discuss some issues of nonlinearity here, in particular how emission reductions are not necessarily linearly related to changes in ambient concentrations. This subject is described well in Chapter 2 of the draft and should be brought forward into the synthesis of key issues with some specific examples of linear and nonlinear relationships between sources and ambient concentrations.

PQ4: "What specific options do we have for fixing the problem? Given the broad control approaches above, what source control alternatives do we have and where can we get the biggest reductions?"

PQ4 is an important question and of great interest to decision-makers. The committee is concerned, however, about NARSTO's ability to provide a comprehensive answer to it. Indeed, the first sentence of the response to PQ4, "Answers to this question go beyond the scope of this assessment," is indicative of NARSTO's limitations. Questions that NARSTO is unable to answer may not belong in this section, although they may be important for pointing out research needs. With that in mind, the committee finds that NARSTO is able to answer the first part of the question for the nine case-study areas described in Chapter 10. The second part of the question, however, is not addressed in the executive summary or in the body of the report, and therefore it may be appropriate to omit it. If NARSTO chooses to answer the second part of PQ4 in this assessment, the case studies should discuss source-control options and provide estimates of the emission reductions expected from each. Such quantitative information is needed for the decision-makers to choose the right policy.

Another problem with the question asked in PQ4 is that it is unclear what is meant by "specific options," as was the case for the term "broad approaches" in PQ3. In a presentation to the committee, Vickery (2002) phrased PQ4 as "What specific *source* options are there?" (italics added). One might expect the answer to this question to include the type of discussion that is found in the third bullet in the response to PQ3. However, the response seems to answer the question "How do we better understand the source contributions?" by giving examples of modeling tools. To alleviate the confusion engendered by the first part of PQ4, the committee recommends that it be reworded to clarify its intent.

Because the current response does not give a clear answer to either part of PQ4, the committee recommends that the response be rewritten to address at least the first part of the question, which NARSTO has already addressed in the body of the report via the nine case studies. A clear indication of specific options available for the case-study areas should be provided by using the text from bullet 3 of PQ3 with recent findings from measurements and models that support the findings. Following the suggested format for responding to all the PQs, the response could then transition into a discussion of the research that would be needed to answer the question better. Recommendations that may be appropriate here include improving the understanding of carbonaceous aerosols, improving emission inventories and models, and conducting case studies similar to the ones discussed in this report for other areas. A shorter version of the wordy and somewhat unclear description of models that currently forms the response to PQ4 could be retained if it included a discussion of how models can be used to reduce uncertainties.

PQ5: "What are the relationships between the PM problem and other problems we are working on, particularly considering sources?"

The committee finds that PQ5, which addresses the interaction between the PM problem and other problems, is appropriate and policy-relevant. However, the question could be rephrased to clarify what types of other problems are of concern and to make the question more consistent with the response. The use of the word "we" contributes to the confusion for this PQ; knowing who is working on the problems could help to clarify what problems are being considered. It appears to the committee that the current response answers the rephrased question "What is the relationship between the PM problem and other problems that the atmospheric-science community is working on?" The subquestions introduce another level of confusion. Because the PM problem is defined for this assessment as exceedance of health-based standards, the questions concerning its relationship to nonhealth effects seem poorly posed. A more appropriate question would be "What is the relationship between PM-management strategies for health reasons and the problems of acid rain and visibility?" Furthermore, the accompanying table addresses the relationship between the PM problem and climate, although climate is not mentioned in the subquestions.

Once the question is clarified, a clear answer should be provided, even if the answer is not yet known for some of the interactions. In fact, the text in Chapters 2 and 5 (referred to in the response to PQ5) indicates that more research is needed to quantify the synergistic and nonlinear effects of copollutants. In the synthesis of key issues discussion, it is important to be clear about what aspects of these relationships are and are not understood.

The current response relies heavily on Table 2.2 to illustrate typical relationships between PM precursors or components and other atmospheric issues. The committee, however, identified a number of problems with the table:

- It is unclear how the indicators were derived, and the level of confidence in each effect is not consistently quantified or communicated.
- Because no column is dedicated to PM mass or to improvements in health, the table does not give sufficient weight to the benefits of reducing each precursor.
- The effect on ultrafine particles is not included as a column under "PM composition," so the table does not indicate the important effect that reductions in black carbon could have on particle size range and thus on visibility and climate change.
- Many of the effects are qualified with *possible* or are indicated as being able to go either way; this makes it difficult to ascertain any definitive understanding of them.
- The color scheme intended to indicate significant expected changes in red, moderate changes in blue, and negligible or unknown effects in black is redundant with the qualifications of many of the effects with words like *possible* and *small*. Without careful reading of the footnotes, a reader may misinterpret the color scheme to indicate desired versus undesired effects, an added piece of information that may be more useful in any case.
- Many of the effects under climate are misleading. For example, saying that a "reduction in NH_3" results in an "increased warming" may be taken to equate reduction of pollutants with increased greenhouse-gas emissions. It would be better to indicate a result of "decreased aerosol induced cooling." In addition, because precipitation and other climate feedback effects are

unknown and could be much greater than the temperature changes, the climate effects could be substantially different from what is noted in the table.

The committee finds that Table 2.2 may be useful as a first step for decision-makers who want to consider tradeoffs associated with particular control strategies. The text should mention that this table is only a first step toward evaluating the risks, and ultimately the costs, of air-quality management decisions. However, it may be prudent to replace the table with a narrative discussing the relationships between PM and the problems of haze, acid deposition, and ozone. The discussion of potential interactions between PM and ozone strategies could draw more from NARSTO's ozone assessment (NARSTO 2000). If the table is retained, the committee recommends that the "Climate Impact" column be removed for the reasons listed above and because NARSTO has not focused on climate science. The discussion in this document should summarize the current Intergovernmental Panel on Climate Change (IPCC) assessment findings pertaining to PM (IPCC 2002).

PQ6: "How can we measure our progress? How can we determine the effectiveness of our actions in bringing about emissions reductions, air quality improvements, and corresponding health improvements?"

The question of how to measure progress is appropriate for this assessment and useful for decision-makers. The question is framed correctly, except that the current response addresses how to measure progress in exposure, a topic that is useful to discuss but one that is not mentioned in the question and is distinct from health improvements.

The response should begin with a clear and succinct summary statement of the present ability to measure progress in emission reductions, ambient concentrations, exposure, human health, and visibility. The summary statement should provide the reader with a general sense of the overall capability of the measurement techniques, monitoring network, and analytic tools available to measure progress. The details of each indicator of progress can be fleshed out in the bullets. The committee finds that this section does not give a fair treatment to the ability to measure progress in health effects. The current discussion makes the prospect of ever measuring progress in PM-induced human health effects sound futile (for example, "Tracking changes in human health effects tied to ambient PM changes is much more difficult, at present is in doubt, and may only be inferred"). Furthermore, the discussion of health in this section should acknowledge that methods are available to relate ambient-air quality to health effects (e.g., Samet et al. 2000) and mention some examples of how reducing exposures has resulted in improved health (e.g., Pope 1989; Chay and Greenstone 1999; Friedman et al. 2001; Avol et al. 2001).

The committee recommends that the response emphasize ways to improve the understanding of how emission reductions lead to changes in ambient concentrations, including any nonlinearities. This subject draws on NARSTO's expertise and is of primary interest to air-quality managers. The current response does not provide much information beyond what decision-makers may already know and what air-quality managers are already implementing. The response could be more helpful to those intended audiences if it commented on what NARSTO sees as the best way to measure progress and key deficiencies in the current system.

PQ7: "When and how should we reassess and update our implementation programs to adjust for any weaknesses in our plan, and take advantage of advances in science and technology?"

The question is policy-relevant and well posed. The aspect of the question pertaining to how often implementation programs should be reassessed is answered directly in the first and fourth bullets, and the other aspects are addressed in the second and third bullets. In general, the committee finds that the response is satisfactory.

The discussion emphasizes how an adaptive approach to PM management that takes advantage of advances in science and technology is essential for building robust policies. Given that premise, the committee concurs with the need for periodic updates of scientific assessments, such as the NARSTO PM assessment. Furthermore, if important new information materializes, such as the identification of particulate species that are most damaging to human health, a reassessment should be conducted sooner than the 6-8 years suggested.

The committee recommends that a figure illustrating the iterative science-policy process be used instead of or in addition to Figure 1.4. Although Figure 1.4 effectively shows the integrative process required for assessing the linkages between pollutant emissions and health effects, it is limited to a scientific framework. Thus, it does not address the question posed in PQ7 and does not support the current discussion of the need for an iterative science-policy process.

PQ8: "What further atmospheric sciences information will be needed in the periodic reviews of our national standards?"

PQ8 is a question that is useful to decision-makers and atmospheric scientists. The response addresses many of the important issues associated with the question. The only large omission is the important role in assessing personal exposure played by study design, in addition to measurement techniques. A schematic showing how the different disciplines interrelate, perhaps building on the framework for informing airborne-PM management and on Figure 1.4, may enhance the presentation. Such a schematic would be especially useful if it identified the gaps in the understanding of each discipline that need to be filled by advances in the others.

Table 8.1, included in the response to PQ8, provides a list of possible harmful components of PM and NARSTO's opinion of the current capabilities for measuring ambient concentrations and personal exposure to those components. Although Table 8.1 is useful for presenting hypotheses that explain why PM is toxic, columns A and B are too abbreviated and therefore at risk of being misinterpreted. For example, a reader might conclude from this table that an objective should be to measure all these aspects of PM routinely. The role of study design in assessing personal exposure is not reflected in Table 8.1. Likewise, the capabilities to measure quantities associated with each hypothesis cannot always be characterized as discretely as they are presented here. An example is in the designation of "research" techniques available to measure ambient concentrations of organic compounds; techniques are available to measure only some of the organic compounds. Indeed, discussion elsewhere in the draft assessment indicates that only a small fraction of the organic mass has been speciated even by the most advanced research methods (e.g., Figure 2.9).

The committee recommends that column B indicating the capabilities of personal-exposure measurements, be eliminated, largely because it does not draw on NARSTO's expertise. Furthermore, the committee finds that the material presented in column A may be better handled in the text, where some of the subtleties could be discussed. Indeed, the only measurements that can be considered fully routine are those of particle mass associated with a particular size cutoff and those of number of particles. Discussing this material in the text could provide an opportunity to indicate what the priorities should be for improvements in measurement technology.

There are two additional ways in which the response could be improved. First, the response calls for improvements in monitoring, emission-inventory development, and modeling, but it does not identify a need for laboratory research. Experiments in the laboratory help to improve the understanding of the chemical and physical mechanisms by which PM evolves in the atmosphere and provide critical input to models; this piece of a comprehensive research program should be mentioned. Second, there could be more discussion of why a mass-based standard may not lead to a reduction in the most toxic components of PM and how atmospheric scientists can assist the health-science community in addressing this topic.

4

COMMENTS ON THE ASSESSMENT'S CHAPTERS AND RELATED APPENDIXES

CHAPTER 1: "PERSPECTIVE AND CONTEXT FOR MANAGING PM"

Chapter 1 introduces the reader to PM, the various factors that affect its composition and spatial distribution, how it affects humans, and what types of regulatory actions have been taken or proposed by the three North American nations. The right assortment of concepts is addressed, but they are sometimes poorly presented. In particular, a reader who is not already well informed on the subject matter would probably carry away incomplete concepts regarding the role of gaseous emission of organic substances, the relative roles and interactions of elemental and organic carbon, the present ability to dissect organic-carbon composition, and the nature and importance of ultrafine PM. Other specific problems are noted in Attachment B of the present report and can be readily clarified by modest editing.

A larger underlying problem with Chapter 1 is that it does not effectively provide a clear framework defining PM and the interconnections between the numerous processes that affect PM and its effects. Readers new to the subject need such a framework to understand PM and how the various chapters in the NARSTO draft assessment fit together. As mentioned in Chapter 2 of the present report, the committee recommends that a general framework for informing airborne PM management, similar to that in Figure 2-1 of this report, be explicitly introduced and defined in Chapter 1 of the NARSTO assessment. The discussion accompanying the figure should explain the processes contained in each box and the interconnections between the boxes. It should also be stated that the general conceptual model presented provides a paradigm of the factors affecting PM but that the complexity of PM makes it impossible to incorporate all the processes and their interactions in a single conceptual model. Indeed, an infinite number of conceptual models could be developed for different regions, times, and conditions, such as those prepared for the nine regions discussed in Chapter 10 of the draft.

Once the framework for informing airborne-PM management is clearly presented in Chapter 1, attention needs to be paid to doing a better job of linking to and being consistent with the rest of the document. Signposts that guide the reader to where in the other chapters one could find more discussion about each topic are essential. The signposts should be clearly associated with the framework (for example, expanded discussion of "Airborne-PM Burden" can be found in Chapter 5 of the draft), thereby enabling readers to understand how the chapters interconnect in this consistently used context.

Chapter 1 is also the place where the PM standards adopted or proposed in the three countries should be presented, especially because these standards are taken as the starting point for the assessment. The current discussion of the standards does not provide enough information about the rationale for the concentrations and averaging times chosen by each country. Ideally, a table that clearly indicates the standards, when they became or will become effective, and whether or when they will be reviewed should be presented. And, it would be useful if some context for comparing standards that have different concentrations, averaging times, and exceedance allowances could be provided.

PM health effects are mentioned briefly in Section 1.4. This discussion would be improved by acknowledging that effects of long-term exposure are perhaps more important than effects of short-term exposure. Section 1.6 very briefly mentions impacts other than health and visibility, but notes that the assessment focuses on those impacts as the primary current drivers of air quality management. Although this limitation of scope may be reasonable, it would provide a useful perspective for readers to expand this section by including at least a paragraph describing other impacts more explicitly and giving one or two key references for each. Such impacts might include those listed in the "Impacts" box of Figure 2-1 of this report. Perspective might also be improved by mentioning that although the health effects are currently the primary driver for concern over PM, it is plausible that other effects, such as climate, might possibly become more significant drivers in the future.

CHAPTER 2: "ATMOSPHERIC AEROSOL PROCESSES: HOW PARTICLES CHANGE WHILE SUSPENDED IN THE AIR"

Chapter 2 is a good tutorial on the chemistry and microphysics of atmospheric aerosols and should be generally accessible to someone who has taken classes in undergraduate chemistry and physics. Jargon is not too dominant and is reasonably well defined when necessary. Nevertheless, a terminology box, as previously suggested for all technical chapters, would be helpful to many potential readers. In addition, there are quite a few awkward sentences (some have been flagged in Attachment B of this report), and the chapter would benefit from a thorough and professional copyediting. As mentioned in Chapter 2 of this report, the recommendation boxes are confusing because they are listed in no apparent order and some numbers are duplicated.

The discussion in Chapter 2 emphasizes science, with the policy implications of the science largely confined to Sections 2.6, 2.8, and 2.9. The policy-relevant points should be summarized and re-emphasized in a concluding policy implications section, as suggested for all technical chapters. Ending the chapter with a section on PM climate effects tends to be a diversion rather than providing a focusing summary.

The draft PM assessment does not describe in enough detail the key role that laboratory experiments play in developing an understanding of the specific chemical and physical mechanisms by which PM evolves in the atmosphere. Chapter 2 is the logical place to provide some discussion of this topic, and the committee recommends that such discussion be included. Several of the recommendations from Chapter 11 that are included in boxes in Chapter 2 (for example, 1.3, dealing with gas-particle conversion effects for organic aerosols and 3.4 dealing with modeling particle microphysics) will not be achievable without substantial innovative experimental laboratory work in heterogeneous chemistry and aerosol microphysics.

The committee notes several problems with Sections 2.8 and 2.9, in which the discussion aims to link the PM problem with other pollutants, visibility, and climate. A number of weaknesses in Table 2.2 were identified in the review of the response to PQ5, where the same table is used. Not only is the table almost impossible to read in the print form, but it is not discussed at all in the Chapter 2 text, despite the clear need to explain it. The table should be either eliminated or better integrated into the discussion. The paragraph in Section 2.8 on PM and haze is both too terse and redundant with Chapter 9, which is not cited for a more thorough discussion. It could be shortened or even deleted with a suitable signpost to Chapter 9.

Climate is addressed briefly in Sections 2.9 and 9.4; neither treatment is sufficient. The climate discussion in Section 2.9 is outdated by the publication of the most recent report of the Intergovernmental Panel on Climate Change (IPCC 2002). The specific forcing values presented should be updated with those in IPCC (2002). The discussion in both sections is a little too simplistic in concluding that the factors governing the direct climate effect and haze are the same. The direct climate effect is actually a balance between visible and near infrared scattering and absorption. Scattering, which leads to cooling, dominates for most aerosol particle types. In contrast, absorption, which leads to heating, can be important for some particle types. The committee recommends that the subject of PM climate effects be addressed well once rather than poorly twice.

CHAPTER 3: "EMISSION INVENTORIES"; APPENDIX A: "EMISSION CALCULATIONS AND INVENTORY LISTINGS"

Chapter 3 summarizes emission information available for Canada, Mexico, and the United States. The types of emission inventories and their uses, uncertainties, and limitations are presented. The chapter also discusses the spatial and temporal patterns of PM emission, and provides PM inventories for various locations and sources. In general, the chapter contains appropriate subject matter. However, one missing topic is the role of transient emissions from off-normal operating conditions (such as, startup of a waste incinerator) in overall emissions and uncertainties in their inventories.

The presentation of material in this chapter is uneven. First, the descriptive text needs a thorough editing to reduce repetition and increase clarity. Second, the tables and the analysis of the data derived from the tables need to be checked for accuracy. Some obvious typographic errors and miscalculations are identified below and in Attachment B of this report. Finally, as discussed in Chapter 2 of this report, the recommendation boxes are confusing. In Chapter 3 of the draft assessment, the placement of the boxes sometimes appears to be arbitrary and has little relation to the surrounding text.

There needs to be a better crosswalk between this chapter and other chapters in the assessment to ensure consistency. The presentation of Mexico City's emissions in Chapter 10 is much better than that in Chapter 3. The information contained in CAM (2001) and Molina and Molina (2002) may be useful for improving the discussion of Mexican emissions. It is unclear whether the research recommendations for emission-inventory improvements presented in Chapter 11 are adequately supported in Chapter 3. The discussion of using receptor-based and source-based models to validate emission inventories in Chapter 3 appears inconsistent with the discussion of these tools in Chapters 6 and 7. Chapters 6 and 7 seem to indicate that the modeling tools can be used in a "weight-of-evidence" mode to help to evaluate the accuracy of inventories. The authors should be consistent as to whether emission inventories can be definitively validated or merely evaluated against the results of source-based and receptor-based models.

The comparison of Canadian and U.S. particulate inventories listed under Table 3.3 would be more informative if numbers were shown as intensities, that is, normalized by population or gross domestic product (GDP). For example, in comparison with Canada, the United States, with 9 times the population and 11 times the GDP, produces only 3 times as much industrial primary PM_{10} or $PM_{2.5}$, 2 times as much industrial SO_2 or VOC, 6 times as much NO_x, and 7 times as much NH_3.

Tables 3.5 and 3.6 should also be normalized by population, number of vehicles, or other parameters to enable comparisons among these locations. This normalization will allow the authors to check whether the data are consistent and accurate. For example, comparing on-road emissions in Tables 3.5 and 3.6 on a per-vehicle basis shows that Los Angeles has much lower SO_2 emissions than Atlanta or Mexico City, as would be expected from the lower sulfur in the fuel. However, the Toronto number appears to be inconsistently low, given the relatively higher concentrations of sulfur in today's Canadian gasoline.

The authors appear to be occasionally confusing the discussion of emission inventories, which are related to sources, with atmospheric concentrations, which is the relevant measure of human exposure. PM concentrations to which humans are exposed are the result of emissions that have undergone some amount of atmospheric processing, which can change the concentration, composition, and spatial distribution of PM. Indeed, airsheds can have widely different assimilation capacities for PM and lead to large differences in exposure for a given magnitude of emission. The differences between emission characteristics and ambient concentrations can be especially large for secondary pollutants and pollutants that can be transported over long distances. The text occasionally gives the impression that ambient concentrations and their resulting exposures can be understood merely through an understanding of emission inventories.

CHAPTER 4: "GAS AND PARTICLE MEASUREMENTS"; APPENDIX B: "MEASUREMENTS"

Overall, Chapter 4 is better written and edited than many others. As is the case with other chapters, a terminology box would help many readers with technical terms. The tone of the introductory discussion (pages 4-1 to 4-4) is appropriate. The various reasons for making measurements, which are not widely known, are accurate, and the discussion of the types of things that need to be measured is clear and useful. Section 4.1, which forms the bulk of the chapter, is an accurate, comprehensive, well-written, and effectively organized description of the current state of research and monitoring instrumentation. Short descriptions of the measurement instruments and techniques for various target data types, and their strengths and weaknesses, are presented in the chapter's subsections; and more detail, suitable for the measurement specialist, is provided in Appendix B. Both the chapter section and the appendix are well organized by measurement type; the two of them combined are a tour de force.

A few overview points, however, are missing from Chapter 4. First, the research-grade instruments and deployment strategies used to gain scientific understanding are often much more sophisticated, accurate, and reliable than the instruments and techniques used to monitor compliance; to develop emission inventories, exposure assessments, and source attributions; and to evaluate program success. Thus, some measurements are of much higher quality than others. That point needs to be made more clearly.

Second, the advancement of key technologies enabling improved instrument design and deployment strategies has been steady. However, for various regulatory, management, and economic reasons, the penetration of advanced measurement techniques into monitoring and compliance applications has been much slower. In fact, many monitoring networks and exposure- and emissions-assessment activities depend on instruments and measurement strategies based on 30-year-old technology. Thus, many of the measurements used to monitor exposure, visibility, and compliance, to develop source attributions, and to evaluate program success are far less comprehensive and reliable than possible. It would be useful to policy-makers to discuss the technologic limitations of current ambient monitoring, exposure and emission assessment, and program evaluation and any programmatic reasons for those limitations.

Third, insufficient attention is paid in the draft assessment to the design of measurement strategies. For example, issues pertaining to the frequency of monitoring, duration of sampling, and number of monitoring sites are not extensively addressed but may be equally important with respect to the specific types of measurements that are made. Better-quality measurements are not necessarily helpful unless they are made in the context of well-designed (ideally, statistically based) networks or studies. In addition, except for the discussion of remote sensing instruments and some terse mention of aircraft-inlet issues, it seems to be assumed that most sampling instruments will be used only at fixed surface sites. However, the development of accurate, real-time (measurement time, of a few seconds or less) instruments opens up opportunities for both airborne instruments (on manned and unmanned aircraft and balloons) and ground-level instruments (on vans, boats, and trains) that can make measurements with better spatial resolution and coverage. The committee suggests that a short section on measurement strategies be added, motivated by the fact that current fixed-site deployments badly undersample the atmosphere. Current measurement data are almost always more sparse than the spatial resolution of analysis and assessment models, and this makes model validation extremely difficult.

The recommendations drawn from Chapter 11 and interspersed in this chapter are sometimes oversimplified and neglect some of the related subtleties and implications. The recommendations are not always fully supported by discussion of where abilities lie now and where they are expected to lie in the near future. Many of the recommended measurement goals will require enhanced versions of existing instruments, such as continuous chemical-speciation measurement and real-time single-particle chemical-speciation techniques. Chapter 5 should identify the research activities necessary to attain the recommended instrument innovations. The implications of the recommendation for more continuous-monitoring development are not fleshed out. Clearly, not all the components that can be measured with filter techniques can now be measured continuously, so replacement of filter-based measurements would result in an increase in temporal resolution at the expense of decreasing the specificity. It is not clear when this is warranted; it certainly depends on the ultimate uses of the measurement data. If the recommendation is not suggesting that real-time measurements of all the species are desirable, those of highest priority should be identified. On the other hand, real-time, hour-by-hour data could be useful to the health community (for example, to address hypotheses of spiked versus integrated exposure) and may be justified on that account.

Section 4.2, on measurement uncertainties, is an excellent overview of individual measurement uncertainties. This topic is often neglected and has policy implications. However, the topic of uncertainties due to undersampling or otherwise inappropriate sampling of the atmospheric system are not mentioned and should be addressed. The section could be improved by adding a paragraph or two at the end to discuss the policy implications of all types of measurement uncertainties.

In general, Chapter 4 could be improved if the policy implications of the scientific and technical information presented were more clearly delineated. The chapter should address how measurements are now used in air-quality management and what advances in measurement technology are necessary for management. Furthermore, the chapter would be more useful to air-quality managers if it provided information for choosing among the different measurement strategies and devices, perhaps via a table that compared the size, sample duration, specificity, sensitivity, and other characteristics of each. Finally, Chapter 4 needs to be clear about the accuracy and precision of each measurement technique and the resulting policy implications. The information presented in Appendix B is useful, but the presentation could be improved. Of particular concern is that the numerous tables are not discussed or referred to in the text. Many tables have titles and captions that are actually discussions that should be in the text. A number of the tables (such as, Table B.1) have several empty boxes; it should be clarified whether this means that the box is "not applicable" or the information required is "not available" or that the answers are "variable" and depend on different researchers.

CHAPTER 5: "SPATIAL AND TEMPORAL CHARACTERIZATION OF PARTICULATE MATTER CONCENTRATION AND COMPOSITION"; APPENDIX C: "MONITORING DATA: AVAILABILITY, LIMITATIONS, AND NETWORK ISSUES"; APPENDIX D: "GLOBAL AEROSOL TRANSPORT"

Chapter 5 is generally well done, presenting a host of PM_{10} and $PM_{2.5}$ data from a wide variety of North American monitoring networks. Recent and historical data from a number of measurement programs are summarized cohesively. These data are the best available representation of the scope and scale of the North American PM problem, including its seasonal variability and 10- to 15-year trends. The chapter is critical in setting up Chapters 6 and 10 and in providing the foundation for the responses to PQ1-PQ3. In short, the data presented in Chapter 5 form the basis for the most policy-relevant part of the report—the regional descriptions of the PM problem presented in Chapter 10. The tie-in between Chapters 5, 6, and 10 should be noted in the Chapter 5 introduction, and the authors should ensure that the discussion in Chapter 5 adequately supports that in Chapters 6 and 10.

Many of the figures presenting time-series data for various monitoring sites or comparing data from multiple sites for the same timeframe are compelling because they illustrate well the high degree of spatial or temporal variability in the aerosol burden. Conveying the highly variable nature of the PM problem in the temporal and spatial dimensions is an important goal for Chapter 5. The data plots are generally effective, although several still fall short of the clarity and readability that are necessary, so it was difficult for the committee to evaluate them. Specific deficiencies are noted in Attachment B of the present report. To be fair, Chapter 5 starts with a note that some of the chapter figures are still under development. If properly done, the figures will be worth many words, so it is critical that they be done well.

Many of the data in Chapter 5 are from the gray literature (publications that are not peer-reviewed, not easily accessible, or both) or personal communications. If the discussion in the chapter is going to rely heavily on gray literature, it is critical to tell the reader how to find the data. And if no widely distributed publication is available, more of the details of the data collection should be included in the chapter (or in an appendix). The most egregious example of this problem is the frequently referred-to Hansen (2000) document, which is a memo rather than a publication and should properly be cited as a personal communication. Because a large fraction of the data presented in this chapter are described in the memo, either it needs to be submitted to a peer-reviewed journal or the data-collection method needs to be detailed in the assessment. Other gray-literature documents that are heavily referred to are Vet et al. (2001), an internal report available on request from the Meteorological Service of Canada, and CARB (2001), a similar internal report. In general, the authors should include references to Web pages where the data can be found and make it clear to the reader that the information is based on work that has not been peer-reviewed.

Chapter 5 does a good job of presenting the variability of the atmospheric PM burden, but it is largely silent on the variability imposed by measurement error in the data presented. There is some useful discussion of the analytic errors that affect data accuracy and the statistical sampling and measurement errors that affect data precision in Sections C.2 and C.3 of the supporting Appendix C. A statement about the general level of accuracy and precision of the data in the chapter needs to be included, and the interested reader should be directed to Appendix C for further information. The information on measurement quality could also be presented in a more coherent and complete manner in the appendix.

Another aspect of variability in PM that Chapter 5 does not address sufficiently is spatial heterogeneity within urban areas. The health community is increasingly concerned about health effects on finer scales than the current monitoring network is able to resolve. For example, recent studies indicate that proximity to roadsides may be associated with PM health effects (e.g.,

Brunekreef et al. 1997); but the current network does not typically include any routine roadside monitoring. The committee recognizes that the assessment authors are limited by the available data but recommends that they augment the discussion of smaller-scale variability and its relation to PM health effects. That can be done, in part, by referring to the discussion in Chapter 8 of the assessment. The authors should also add discussion of how the atmospheric-science community might address this outstanding research question better.

Mexico is not satisfactorily represented in Chapter 5. Just as a key PM-problem region crossing the U.S.-Canada border is identified in the Pacific Northwest, several areas with severe PM problems cross the U.S.-Mexico border. Those areas are the focus of much current research because expected human exposures to PM are probably very high. Many factors contribute to the high levels of particle concentrations that have been observed in those areas, including the nature and characteristics of the emission sources, land use and topographic characteristics, and meteorologic conditions (Mukerjee 2001; Mejia-Velazquez and Rodriguez-Gallegos 1997). Chapter 5 should discuss the areas. In addition, the figures in Chapter 5 that already include Mexican data often are inadequate (see Attachment B of the present report). A better attempt could be made to discuss Mexico's PM characteristics more completely.

The introduction of Chapter 5 tries to set up the theme of widely varied spatial and time scales and their effect on PM distributions and trends. The text is reasonably successful, but Table 5.1, which is supposed to summarize the message, is unsuccessful. First, the "global" scale does not start at 1000 km; in fact, for tropospheric aerosol, it is debatable whether there are any truly global effects—even aerosol climate forcing operates on a continental or ocean-basin scale that is semiglobal at most. Second, and more serious, by apparently mixing the time scales for aerosol processes with the time scales for such aerosol effects as health effects, which can be acute (hours to days) or chronic (years to decades), the table obscures all sense of the connection between atmospheric-process spatial and time scales. Table 5.1 needs much more thought so that it will support rather than apparently contradict or confound, the text of Section 5.1.

Section 5.2, "The Influence of Global Aerosol Transport on PM Mass Concentrations in North America," is poorly positioned in the chapter in that the data show that global transport is not significant. Although this section is scientifically accurate and interesting, starting the chapter by discussing an insignificant determinant of observed aerosol variability will draw the reader away from factors that are important. The committee suggests shortening the section and making it a subsection with the same title right after the current subsection 5.4.4., "Regional Transport." As noted above, transport of aerosols is still at most semiglobal or supraregional; dust can be transported across an ocean but generally not around the entire globe.

Most of Section 5.4, "Regional and Urban Contributions to Particulate Matter," is well motivated and well done. The discussion of weekly cycles is interesting, especially the difference between $PM_{2.5}$ and PM_{10}. The data showing the weekly pattern is potentially powerful and needs to include references to published work to allow the reader to be convinced that it is statistically relevant. The section on visibility in the Grand Canyon is redundant with, and in some cases contradicts, material on the same topic in Chapters 6 and 9. The committee strongly suggests integrating the material from page 5-37, line 31, through 5-39, line 10, into the relevant material in Chapter 9. Chapter 5 can mention the regional-transport evidence from the Grand Canyon study in one short paragraph and refer the reader to Chapter 9 for details, thus reducing a serious redundancy in the overall assessment. Furthermore, special attention should be paid to making all discussion of the Grand Canyon study consistent.

Section 5.6, "Covariation of PM with Ozone and Other Air Pollutants," successfully makes the important point that ozone correlates with only part of the PM burden—and then only part of the time. However, it could be made more succinctly but just as forcefully. In addition, the only data presented compare ozone with various PM measures, so the "other pollutants" in the section title is misleading. In general, this section could use some copyediting to tighten the text.

The current "Conclusion" section is done well from a scientific-summary point of view. However, it begs to be rewritten so that its many important policy implications are explicitly identified. As is the case with the other technical chapters, the committee recommends that this section be rewritten in this fashion and relabeled "Policy Implications." In particular, the policy implications of Section 5.5, "Trends and Their Implications," one of the most policy-relevant sections in the assessment, should be more definitively expressed here, and another signpost to Chapter 10, where many of the policy implications are further illuminated, should be planted.

CHAPTER 6: "RECEPTOR METHODS FOR SOURCE APPORTIONMENT—BEYOND THE EMISSION INVENTORIES"

Chapter 6 presents a comprehensive review of receptor-modeling methods. The authors have done an excellent job of reviewing the science. In particular, they have tied together disparate approaches that have been developed over the years and have explained the basic concepts in lay terms while providing some seminal references for the reader. Nonetheless, a number of issues should be improved before publication.

Better coordination between Chapters 6 and 7 is needed. A consistent approach to framing and contextualizing Chapters 6 and 7 would enhance the presentation particularly because the two categories of tools are complementary and can be used to address overlapping sets of questions. Furthermore, the two chapters would benefit from being more consistent with the rest of the assessment. The committee recommends that each chapter begin by explaining how the receptor- or source-based modeling tools fit into the framework for airborne-PM management, which was ideally presented in Chapter 1. How the tools in Chapters 6 and 7 are similar and different should also be explicitly described.

Each chapter should present the policy relevance of the material in a parallel manner in a final "Policy Implications" section, as recommended for all technical chapters. The policy-relevant material in Chapter 6 is now found in Section 6.5, which addresses how modeling tools are applied in PM policy development, and in Section 6.6, which includes answers to "science questions" that are similar to but not exactly the same as the PQs used in the executive summary.

Chapter 6 would benefit from heavy editing to eliminate sections that are presented and discussed elsewhere and to improve the consistency between it and the rest of the assessment. There seems to be substantial overlap with other chapters, including the discussions of single-particle mass spectrometers and of haze in the Grand Canyon. The definition of manageable and unmanageable source contributions is introduced more successfully in Chapter 1 than in Chapter 6. Definitions of terms in Chapter 6 are not entirely consistent and should be better aligned with those in Chapter 1. For example, in Chapter 6, "Saharan dust" is listed as an unmanageable source, but "wind-blown dust from natural areas (e.g., deserts) . . . that is accentuated due to land-use or industrial practices is at least partially manageable." Likewise, the idea of "weight-of-evidence," introduced on page 6-49 as the "agreement between receptor and source models," needs to be treated consistently and introduced in Chapter 1 of the assessment. The only previous mention of weight-of-evidence is in the executive summary, where it is defined as the "use of integrated information from emissions inventories, ambient concentration measurements, and air quality models." The idea of a conceptual model is defined and used in the most detail in Chapter 6, but it still is not very clear. As discussed at length in Chapter 2 of the present report, the committee recommends that the treatment of conceptual models be consistent throughout the assessment.

Despite the problems with repetitiveness and consistency, Chapter 6 is very readable, although possibly at the expense of a number of details and recent results that would provide more specific information. For example, the discussion of source tracers is informative but would benefit from being more quantitative. Such species as vanadium are good for identifying some fossil-fuel combustion sources, but their utility often depends on measuring a sufficient quantity relative to other species to obtain a clear signal. In addition, although some sources have a number of elemental tracers, not all have them in sufficient quantities to be useful for identification. The graphs in Figure 6.5 illustrating several source signatures are useful for making that point, but the discussion in the text needs to link to the figures better. Another example is the diesel-emissions discussion (page 6-39), which neglects the possibility that these emissions may produce ultrafine particles not captured by $PM_{2.5}$ mass measurements at very high number concentrations or fluxes. Although few measurements of ultrafine particles exist, it would be worth pointing out the potential health effects associated with them, especially considering that other currently unregulated factors (for example, cold starts and vehicle maintenance) have been discussed.

A number of the examples chosen to illustrate points made in Chapter 6 are not appropriate or effective, possibly because not many case studies that assess PM-management strategies are available. The discussion of sulfur reductions in Canadian gasoline, which is intended to give an example of how to use a conceptual model and receptor-oriented methods to identify emission-reduction targets, is par-ticularly problematic. It is a poor example of PM management because fuel sulfur reductions were not intended to reduce primary emission of PM. In this example, the chemical mass balance model indi-cates that vehicle emissions are responsible for 50% of $PM_{2.5}$ in Toronto and 42% of $PM_{2.5}$ in the Lower Fraser Valley of British Columbia. The motor-vehicle source is not speciated in Figure 6.7, so it is unclear whether the $PM_{2.5}$ from motor vehicles is due mainly to sulfates, organic compounds, or nitrates. The accompanying text indicates that reducing emissions of fuel sulfur by more than a factor of 20 was estimated to decrease $PM_{2.5}$ by only 1.8%. That ambient $PM_{2.5}$ was predicted to be largely unaffected by fuel sulfur reductions suggests that perhaps other sources might be better emission-reduc-tions targets. It seems that one could conclude from the modeling activities that targeting the large amount of nonvehicular ammonium sulfate or vehicular emissions of organic material and nitrate may be more effective for reducing $PM_{2.5}$. One might even conclude that reducing fuel sulfur is not an effective strategy for PM control. It would be better to choose an example in which the receptor-based modeling methods are used to reduce $PM_{2.5}$ more dramatically.

Further muddling that example is Figure 6.8, which is confusing despite seeming to be central to the discussion. It is unclear how to read the diagram: Does the bottom right corner indicate that changing fuel causes emission changes in the top left? If so, the directionality of the arrows and having some of the arrows go in two directions do not make sense. And, the intent of the boxes labeled "Approach . . . " and "Responds . . ." is unclear. For example, some boxes under the "Approach" label appear to indicate emissions. The discussion of Figure 6.8 refers to colored and numbered boxes, but the figure is not in color, and no boxes are numbered.

A second problematic example is the discussion of "Haze in the Grand Canyon." The modeling activities indicated that emissions of SO_2 from the Mohave Power Project did not contribute much to the sulfate observed over the Grand Canyon, but a decision to install sulfur scrubbers was made nonetheless. Again, the modeling results did not seem to guide the policy decision. The comment on page 6-47 about how the results "were most influential in determining the timing of implementation rather than whether or not to implement sulfur reductions" is confusing. Were the pre-existing plans to install sulfur scrubbers delayed because the modeling results indicated that the effect would be small? Perhaps the example should be presented as a study that should have been done before the decision to implement scrubbers was made. As it stands, the authors seem to have concluded that the money spent on installing scrubbers will have a negligible effect on visibility and that the investment in the scrubbers will be shown to have been an unwise policy decision. If that was not the point of the example, it needs to be better clarified in the text. Understanding the example is complicated by the confusing nature of Figure 6.11 and its caption, which together provide the basis for the assertion that the receptor models indicated that sulfur from the Mohave Power Project contributed only a small fraction to sulfate over the Grand Canyon. If that is the key result of the study, the figure and the caption should be made abundantly clear, especially because not all readers will be familiar with cumulative-frequency plots; for example, it may be unnecessary to include the left panel for this purpose.

CHAPTER 7: "USING MODELS TO ESTIMATE PARTICLE CONCENTRATIONS AND EXPOSURE"

The role of chemical transport models (CTMs) in calculating particle concentrations and exposure is addressed in Chapter 7. Overall, this chapter is well written and comprehensive and constitutes a useful summary. Relative to the other chapters, Chapter 7 has a large number of specific references. That is not totally consistent with the style of the other chapters, but it is a better model, and the other chapters could include more specific citations.

As discussed in the comments on Chapter 6, Chapters 6 and 7 should be framed more consistently to draw more attention to how the tools presented in each chapter complement each other. Chapter 7 should be made more consistent with the rest of the assessment, via the framework for airborne-PM management and a final "Policy Implications" section, as recommended for all technical chapters. Because CTMs try to represent all the factors influencing PM discussed in previous chapters of the assessment (such as atmospheric aerosol processes and emissions), the authors should ensure that the discussion in Chapter 7 is consistent with previous treatments of the topics and try to avoid repetition. Likewise, how CTMs are used to elucidate the aspects of the framework for airborne-PM management, particularly the conceptual model for airborne-PM burden and the management scenarios, should be described.

The policy-relevant material in Chapter 7 is in Section 7.4, "What Questions Can Chemical Transport Models Address and How Well?"; Section 7.9, "Policy-relevant results from CTM applications"; and Section 7.10, "Critical Uncertainties." The summary in Section 7.11 is adequate from a technical perspective but could do a better job of highlighting policy-relevant contributions. For example, the discussion of uncertainties is useful for guiding policy, but that connection is not made in this section. Therefore, a concluding section should be added that lists the highest uncertainties that policy-makers could address by requiring specific measurements.

Chapter 7 includes many references to uncertainties associated with CTMs, both in Section 7.10, "Critical Uncertainties," and in the sections preceding it. To provide effective guidance to decision-makers, it is important to provide, if possible, some quantification or at least qualification of the uncertainty. Indeed, the authors should ensure that a reader does not conclude that CTM results should be discarded altogether. It should be made clear in the introductory material that CTMs will never be able to represent all the processes in the atmosphere precisely and that approximations are therefore necessary and often reasonable if based on good assumptions and data. Ideally, the discussion should relate clearly how CTMs can be effective tools for air-quality management despite their limitations.

The important point needs to be emphasized that although CTMs function reasonably well under normal, or the most typical, meteorologic conditions (that is, most of the time, in most places), they do not function well under unusual meteorologic conditions. Not only might the meteorological model be unable to reproduce unusual conditions, but model components which use meteorological data as input (for example, chemistry modules use temperature and solar intensity) may include approximations or assumptions that do not apply under extreme meteorological conditions. Because pollution episodes often occur as a result of unusual meteorological conditions, models may have limited ability to predict such episodes. That point could be reinforced in both the sections describing the models and in the summary.

It would provide a helpful context to include more information on the accuracy of model predictions. The treatment of this topic in Section 7.6, "Current Status of CTM Performance and Intercomparisons," is meager. The only measures of actual model performance given are in Table 7.1, a rather dense summary of model evaluation that is not immediately accessible to an uninformed reader. The section would benefit from figures giving specific examples of how model results compare with observations. The summary record (or at least examples) of the accuracy of predictions of both episodic and longer-term averaged PM concentrations (or those of other pollutants if there are no PM data) would be useful to portray the state of the science (and art).

The model intercomparison in Table 7.1 is limited by the fact that the models were not all run for the same situation or with common inputs. The statistics used to compare them are not even uniform. Some discussion of the prerequisite conditions for meaningful model comparisons should be included. For models to be compared effectively, they would ideally simulate the same episode, have identical inputs for emissions and meteorology, use common formulations for basic physical and chemical processes, have a standard set of rate constants, and produce output in the same statistical format.

The extent of use and congruence of CTMs among the three nations should be discussed. Are CTMs primarily research tools, or are they being used routinely for policy-evaluation studies (and are there differences among the countries)? To what degree is there uniformity of models and model use among the countries? If they are not standardized, what advantage, if any, would there be if models for specific tasks were standardized among the countries? Should the atmospheric-science community be trying to develop standardized process submodels or input datasets (as the stratospheric-chemistry community has done)?

Chapter 7 does not give a balanced account of how CTMs can be used to estimate exposure. The difference between the information needed to estimate concentrations and exposure is not recognized in most of the chapter; in fact, Figures 7.2 and 7.3 deal only with the steps for obtaining concentrations. If estimating exposure is intended to be a central topic in Chapter 7 (as the title indicates), discussion of it should not be relegated to vague comments in Section 7.8. Figures 7.2 and 7.3 should be modified to show the additional inputs and steps needed to determine exposure on the basis of concentration (regardless of whether they have been implemented). Furthermore, the discussion in Section 7.8 should be more specific and provide examples of exposure estimates. Since no real analysis of exposure to ambient PM is provided, the authors may want to remove exposure from the chapter title.

CHAPTER 8: "HEALTH EFFECTS OF PARTICULATE MATTER"

Chapter 8 is the most problematic chapter in the draft assessment. The committee debated the utility of retaining this chapter in the assessment or of starting with the assumption that sufficient health "drivers" exist and referring in the introduction to other summaries of PM health issues (e.g., NRC 1998, 1999, 2001; EPA 2002; Molina and Molina 2002). Removing the chapter, however, would eliminate the little emphasis on health now included in the assessment, detract from meeting objective 5, and contradict assumption 2 as listed in the NARSTO charge (reproduced in Box 1-1 of the present report). Indeed, the committee feels that a chapter devoted to health effects of PM and opportunities for linking health and atmospheric-science research could be very useful, if written well, and recommends that a chapter discussing PM health effects be retained.

The committee recommends that Chapter 8 undergo major revision to correct conceptual errors, improve readability, reduce reliance on technical terminology, and highlight opportunities for interaction between health and atmospheric scientists. Ideally, this chapter should provide atmospheric scientists with a highly accessible summary of PM health effects that helps them to frame their work better. References to more-detailed assessments should be provided. The summary of PM health effects should also be highly accessible to decision-makers; this assessment may be their only exposure to the topic, and it is critical that the summary be conceptually correct.

In addition to providing a current, readable summary of the health effects of PM, the committee recommends that Chapter 8 be rewritten to focus more on the interface between atmospheric and health sciences, on how to facilitate flow of information between the communities in both directions, and on any resulting policy implications. In particular, it would be highly useful to discuss information known or being investigated by the health community that would help atmospheric scientists. Likewise, a discussion of research avenues for atmospheric scientists to pursue to assist health scientists would be useful. For example, epidemiologic analyses indicate spatial heterogeneity in PM health effects, but the extent to which the spatial variability can be explained by PM composition or atmospheric chemistry is unknown. As another example, the importance of understanding exposure to a putative agent (for example, PM or a sub-fraction of PM) for confirming the relationship between a health outcome and that agent points toward the need for coordinated involvement of atmospheric and exposure scientists.

Numerous specific corrections for Chapter 8 are provided in Attachment B of the present report. In general, the material is too technical and specific for a target audience of air-quality managers, policy-makers, and atmospheric scientists. Many in the intended audience are not likely to understand some of the terms without further explanation, such as *spline curves*, *lag*, *interquartile*, *log of mortality-days*, and *Akaiken Information Criteria*. The chapter should be written so that readers only modestly informed about health issues and the technical aspects of dosimetry and risk estimation can understand it. It should focus on selected key concepts and avoid being an abstract of technical information. In addition, the presentation is uneven: the personal-exposure section needs substantial editing, has several factual errors, and is out of date in places; the epidemiology section is in better shape.

Chapter 8 needs editing to provide the reader with an appropriately broad and accurate overview of health issues. The approach of paraphrasing and excerpting text from the U.S. Environmental Protection Agency (EPA) criteria document for PM (EPA 2002), which is intended to be a comprehensive survey of the health-effects literature, is not the best for the NARSTO assessment. Chapter 8 relies so heavily on the EPA criteria document that the reader would become much better informed by reading summary sections of that document. Indeed, the reader may not be able to grasp the information in this chapter without having the EPA document close at hand and referring to it often to place information in context, get more background on a subject, or read a better explanation.

Many recommendations included in Chapter 8 are not specific and therefore not very useful—for example, "Comprehensive, directed study of the factors [that influence exposure] would also be of tremendous value" (page 8-5) and "All of the epidemiological studies of PM and other air pollutants will benefit from close collaboration between atmospheric and health scientists" (page 8-12). In several places, the chapter notes that closer interaction between the health-science and atmospheric-science communities is needed but gives no specific examples of beneficial interactions. The committee agrees that there ought to be more communication between the communities, but mentioning the issue is just lip service unless the nature and potential results of specific types of interactions are described.

CHAPTER 9: "VISIBILITY EFFECTS"

Chapter 9 gives the reader a basic understanding of visibility impairment. The committee notes that visibility and regional haze are also discussed in Chapters 1, 2, 5, and 6. Efforts should be made to reduce repetition by concentrating the discussion in Chapter 9, having other chapters refer to the discussion in Chapter 9, and ensuring that all treatment of the topic is consistent. Of particular importance is a consistent treatment of the roles of light absorption and scattering and of how water uptake by PM affects visibility.

The issue of managing visibility is taken up in several places in Chapter 9, including Sections 9.1.4, 9.1.5, and 9.5. The committee recommends that the chapter be reorganized to address management in a single section. Current or proposed rules, regulations, and programs that address visibility, which are absent from the draft, should be introduced and described in this section. To the extent practicable, the text should address how and why each standard was set, what measurements they are based on and how the measurements are obtained, and whether the standards have been effective in improving visibility. Some discussion of the recently proposed regional haze rules in the United States should be included. The discussion in Section 9.5, which logically would be in the new section, would also be improved if more specific information—including details about the modeling tools, control strategies, or lessons learned—were given about the various examples (such as the Grand Canyon Visibility Transport Commission and the Southern Appalachian Mountains Initiative). At least, references should be provided to guide the interested reader to such detailed information. Finally, the new section devoted to visibility management should address the economic and societal effects of visibility impairment, particularly explaining why this effect of PM is of concern.

The first section of Chapter 9 introduces the topic by describing how visibility is linked to PM. Extensive data are presented for the United States, but there is not much information on Mexico or Canada, even though some such information is available. For example, there is some evidence that visibility has decreased markedly in Mexico City from an average of 4-10 km in 1940 to an average of 1-2 km in recent years (Ciudad de Mexico 1999).

Sections 9.1.2 and 9.1.3 describe many of the technical relationships between particles and visibility. How particle composition may increase or decrease visibility is discussed, as is how relative humidity may affect visibility by increasing scattering efficiency. Relationships among PM concentrations, PM type, measurement methods, humidity, and visibility could be presented in more depth. For example, the present discussion does not indicate that measurements of PM on filters, which are dried to some moisture specification before weighing, will not necessarily give an accurate indication of visibility, which is strongly influenced by water uptake.

Section 9.1.4, "What are some unique aspects of visibility management?", needs to be revised or integrated into other sections of the chapter, such as the proposed section dedicated to visibility management. It is not evident from the points presented why visibility management is unique. For example, the authors mention that "visibility is perceived instantaneously, so there is no averaging time." Does that mean that there is no standard or that visibility cannot be modeled (which is not correct), or are the authors trying to make some other point? A good reason to integrate the material in this section into other parts of Chapter 9 is to reduce redundancy. Much of the information presented in the previous section is restated here, in particular that composition, size, and humidity are factors that may modify the effect of PM on visibility. Other points made here fall nicely into the discussion of visibility management, including comments on how visibility is perceived instantaneously and the implications for visibility management in clean and dirty areas. The summary nature of some of the comments also makes them appropriate for a final section of the chapter in which the policy implications of the scientific information are described (as has been suggested previously for all chapters).

Sections 9.2.1 and 9.2.2 provide a summary of the different visibility inventories. This section would benefit if the authors provided references for the studies mentioned. Section 9.2.3 presents a case study of the Colorado Plateau. Although the section is interesting, why this case is presented as a conceptual model and what can be learned from it are not clear. Are there lessons from this case study for air-pollution control offices in Mexico or Canada? Section 9.2.4 poses an important question about using PM studies designed to address health questions to learn about visibility, but the response to it is not complete. The authors should flesh out the response by discussing, for example, the additional benefit of speciation data for understanding the visibility problem in rural, scenic areas and in urban areas.

Section 9.3 starts with a long introduction regarding models available to estimate visibility. The authors should attempt to condense this section, focusing on describing the available models, the information they can provide, measures of their performance, and how they can be used in visibility management. A table that summarizes that information and advantages and disadvantages of each modeling tool may be a more efficient way to present the material. The authors should keep in mind the audience of visibility managers who need to use and develop models to assess the effects of distant sources of PM; in this context, references to specific software and documentation would be useful.

At first glance, Section 9.4, "Atmospheric Aerosols Affect the Global Radiation Balance," seems out of place in a chapter titled "Visibility Effects." However, the discussion about how efforts to understand the role of PM in climate and visibility complement each other and how mitigation strategies could be developed to address both problems is interesting and appropriate for this chapter. The subject is more logically probably a subsection of the proposed section on visibility management. Indeed, the existing Section 9.5 has a parallel discussion of aligning visibility and health-based PM-control programs.

CHAPTER 10: "CONCEPTUAL DESCRIPTIONS OF PM FOR NORTH AMERICAN REGIONS"; APPENDIX E: "CONCEPTUAL DESCRIPTIONS OF SELECTED NORTH AMERICAN SITES"

Chapter 10 offers a summary of the PM problem in nine regions in North America and comprehensive descriptions of the PM problem in the San Joaquin Valley of California, the Windsor-Quebec City Corridor, and Mexico City. Appendix E provides details on the other six regions. The material presented in the chapter is the most original contribution of the PM assessment and provides the most useful information for decision-makers. In particular, the information in the tables provides an excellent summary and deserves greater prominence in the assessment because it clearly links recent research to policy suggestions in a concise way. For those reasons, the committee recommends that all nine of the regional descriptions be presented in Chapter 10 rather than having six relegated to Appendix E. In general, the approach is clearly explained, and the chapter is well written. The committee, however, has identified a number of ways in which this policy-relevant information could be better communicated.

As discussed previously in Chapter 2 of the present report, the conceptual model discussion in Chapter 10 of the draft assessment is not consistent with the other chapters. The committee recommends that the same framework for informing airborne-PM management (Figure 2-1 of the present report) be applied here but in a region-specific manner. Using a consistent definition of the framework and the conceptual model that is part of it would alleviate the confusion introduced by having a "schematic diagram of a conceptual model" presented in Figure 10.1, a somewhat different conceptual model identified in Figure 10.2 (note that the processes included in the conceptual model in Figure 10.1 are inputs to the conceptual model in Figure 10.2), and a third variation of the conceptual model for the nine regional descriptions.

Presenting the nine region-specific conceptual models in the same figure format as the framework for informing airborne-PM management, but with information in each box reflecting the regional characteristics, may be a more effective way to communicate a coherent story about each region. The figures could replace or augment Tables 10.1, 10.2, 10.3, 10.4, and 10.6, which are tedious and informal in their current form (and contain a number of grammatical and spelling errors). The tables map fairly easily into the conceptual-model diagram in Figure 2-1 of the present report:

- Tables 10.1 and 10.2 go into the "Airborne-PM Burden" box.
- Table 10.3 goes into the "Meteorology" box.
- Table 10.5 goes into the "Source attributions" and "Emissions" boxes.
- Table 10.6 goes primarily into "Policy Implications"; some information goes into the "Atmospheric Processing" box.

A second problem with the presentation in Chapter 10 is that the discussion in the text needs to support the tables better. There should be a discussion of how the source attributions (Table 10.5) and the policy implications (Table 10.6) were determined from the information in Tables 10.1-10.3. In addition, the nine sections titled "Implications for Policy Makers" in Chapter 10 and Appendix E should be rewritten to mirror and augment the equivalent sections in Table 10.6. In rewriting those sections, the authors should keep in mind that decision-makers are typically looking for scientists to provide a range of options for addressing the policy issue and, wherever possible, articulate the respective strengths and weaknesses of competing options. By and large, that has not been done in Chapter 10, and the authors should attempt to provide at least some such material.

In discussing policy implications, the authors should carefully consider how they present information that is highly uncertain or demonstrates adverse collateral effects associated with particular control strategies. An example of a discussion of uncertain information can be found in Section 10.4.6, where the emission inventory for Mexico City is said to be "highly uncertain." No comment is made in the section about the importance of reducing the uncertainty, what specifically is uncertain about the inventory, or what causes the uncertainty. Without such additional discussion, a policy-maker may be inclined to ignore any policy implications stemming from the uncertain information. Likewise, wherever tradeoffs are identified in the analysis of possible actions (for example, where solving one problem will or may exacerbate another), it is important to provide further guidance to the policy-maker on the implications of action and inaction in the face of such tradeoffs. For example:

- Are the two problems of comparable significance or magnitude? Why or why not?
- Is there a net benefit in risk reduction of acting one way or the other?
- Are there other aspects to each of the two problems (such as effects on sensitive populations) that would affect how they are balanced against each other?
- Is the tradeoff recommended on cost-effectiveness grounds in any case?

In short, it is critical to provide information that puts the tradeoffs into context.

A final aspect of Chapter 10 that might compromise its ability to speak to decision-makers is associated with the five questions used for organizing the regional descriptions. The five questions bear some resemblance to the eight policy questions (PQs) presented in the draft assessment's executive summary. However, they are not identical and are trying to elicit different responses; the eight PQs are intended to provide a general understanding of PM and its management, whereas the set of five questions in Chapter 10 are intended to provide an organizing framework for discussing the regional descriptions. The wording should be modified to distinguish the objectives of the two sets of organizing questions. Furthermore, the five questions do not exactly align with the five tables presented in Chapter 10 or with how each region is described in the text. The regional descriptions should be presented in a consistent manner within Chapter 10. For example, the section headings should not vary among regions.

CHAPTER 11: "RECOMMENDATIONS"

Chapter 11 does a good job of presenting a detailed discussion of the high-priority scientific research that should be undertaken to improve the understanding of PM in North America. The scientific targets of the research are described in detail elsewhere in the draft assessment, the recommendations are specific and accompanied by their rationale, and they focus on questions that are clearly important for policy decisions and uncertain given current findings. The link of each scientific subject to policy is discussed, but these sections generally need to give decision-makers more rationale for why investments will give the most benefit in addressing the difficulties in making policy choices. It would also be useful for the decision-maker community to identify what aspects of the recommendations are practical, on the basis of current abilities, and on which fronts progress is expected in the near term or the long term.

The committee finds that the recommendations are generally appropriate, but it also feels that the context and framing of the recommendations could be improved. The introductory comments should mention that advocates on all sides of the environmental-science debate agree that good decisions should be informed by a sound understanding of science and its uncertainties, and these recommendations will allow the uncertainties to be reduced. The title and introductory discussion should better reflect the thrust of the discussion in the chapter, that is, what improvements in data, tools, and knowledge are needed for a better assessment of the science and better answers to the policy questions in the future. Indeed, the recommendations of most interest for air-quality policy are in Chapter 10. A more specific chapter title would make it clear that the recommendations pertain only to scientific research that should be pursued, not to PM-management options. The committee recommends that the title of Chapter 11 be changed to

"Future Research Directions," "Science to Inform Future Policy," or something similar of NARSTO's choosing.

In the discussion of research needs, it would be useful to call attention to the fact that some gaps in scientific knowledge for policy-making are due to true uncertainty (for example, magnitude of a local source's contribution to ambient-PM concentrations at a specific location). However, other knowledge gaps faced by decision-makers are due to variability (for example, differences in the susceptibility of individuals to similar particle exposures). These general concepts and terms are discussed extensively in the National Research Council report *Science and Judgement in Risk Assessment* (NRC 1994). That report discusses *uncertainty* in terms of knowledge that does not extend beyond a certain degree of precision because of measurement or estimation error. On the other hand, *variability* is an attribute of a factor, such as regional weather patterns or human characteristics, that does not allow the factor to be represented by a single value. Both uncertainty and variability pertain to issues discussed in the NARSTO assessment, and a clearer exposition of their differences would facilitate a better and more realistic understanding of the issues for deicision-makers.

To frame Chapter 11 better with respect to the material presented in other chapters, the authors should explicitly relate the recommendations to the policy questions, the policy-implications sections to be added at the ends of other chapters, and the limitations identified in Chapter 10. A table may be an effective way to present such information, or it should be addressed in the short "policy-relevance" paragraphs. Those paragraphs currently provide the rationale for the recommendations. They should also indicate how the recommended research would affect the decision-maker's choices. An example of where this link is made clear is the recommendation about carbonaceous aerosols, which states that a better understanding of organic compounds will help in deciding whether reductions in volatile organic chemicals will be effective in reducing PM mass.

There needs to be attention to articulating summary statements and expressing research needs as clearly as possible. Granted, many of the statements are intentionally brief and in bullet form (especially in the tables), but a recommendation is not effective if the reader cannot readily grasp what it means. Many readers will read Chapter 11 without having read the preceding chapters, so it needs to be able to stand largely on its own. Likewise, many readers may read the tables without reading the text, but the concepts are much more clearly conveyed in the text than in the tables. The explicit intent of the summary recommendations in the tables also needs to be understandable, even though not all the details surrounding them can be conveyed there.

The term harmonization is used often in Chapter 11, but it is ambiguous and overused. In some instances, the authors seem to be calling for standardization (that is, everyone using the same approach or model). *Harmony* does not mean everyone doing the same thing; rather, it means different people doing different things that are complementary and that make a whole that is greater than the sum of the parts (as in musical harmony). In a scientific context, the term could refer to two adjacent measurements that may not match exactly but differ in an explainable manner. Or the term could refer to standards that differ but do not conflict. To alleviate the confusion, the wording could be made more explicit when the term is used and examples of harmonization could be given.

An important research need seems to have been overlooked. A much better understanding of PM surface chemistry and morphology is needed. Unless the PM is soluble, cells are not sensitive to bulk chemistry; they are affected mostly by surface chemistry. The atmospheric-science community has made progress in analyzing the bulk composition of PM, either by using collected samples or single-particle analysis. However, little (if any) progress is being made in the atmospheric-science or health-science community in understanding PM surface composition or interactions between PM surface composition and biologic fluids.

Whether or not a specific priority-setting was intended, the order of the five recommendations and their subrecommendations will suggest priorities to the reader. Thus, the order is important and should be the same everywhere it is discussed (that is, on page 11-1, in the tables, in Section 11.1, and in the synthesis of key issues). Likewise, the same wording should be used everywhere the recommendations are presented. Because priorities are useful for decision-makers, the committee recommends that NARSTO attempt to assign priorities to the recommendations for future research in terms of when each activity needs to be undertaken and what should receive more funding. The committee recognizes the difficulty in assigning uniform priorities for all countries and situations. Thus, based on the basis of the level of understanding in different areas, regional priorities may be appropriate.

The committee also identified a number of specific problems with the recommendations themselves. Recommendation 2.8 should mention the relationships between *sources* and health effects, rather than only constituents and health effects. A recommendation for continued and enhanced laboratory studies of aerosol atmospheric chemistry and microphysics is missing. And there is no recommendation for a better understanding of PM surface area and surface composition, which may be more closely associated with health effects.

5

RECOMMENDATIONS FOR FUTURE NARSTO ASSESSMENTS

NARSTO indicates in the draft assessment that it intends to repeat the assessment process at regular intervals, with a second assessment on PM to be conducted in 2007-2008. The committee recommends that NARSTO, in preparing for and conducting these future assessments, focus on enhancing its interaction with the policy community to make it possible for NARSTO to inform policy decisions better. In addition, NARSTO should strive to better place atmospheric-science information in the context of effects, including those on health, ecosystems, and global climate, as well as economics and other social sciences. With stronger links to those communities, NARSTO could be much more effective in communicating its science to researchers in the related fields and in identifying high-impact research directions for atmospheric science.

As discussed previously in this review, the committee finds that the current draft of the PM assessment falls short in the process and methods used to elicit information needs from decision-makers and in communicating the policy implications of the science to them. The committee strongly recommends that social scientists with expertise in elicitation of information be engaged in developing policy guidance for future assessments. Future NARSTO efforts to elicit decision-maker needs and concerns must be designed to use valid techniques to obtain representative and unbiased samples. Furthermore, to meet its goals of aiding air-quality policy and management decision-making (see Box 1-1 in the present report), NARSTO needs to refine how it attempts to communicate scientific information to policy-makers, particularly by drawing bolder and less ambiguous policy implications. Of greatest importance is more explicitly identifying air-quality management activities that will most likely lead to the desired outcome; tradeoffs, nonlinearities, and synergies associated with those activities; and alternative options. Those are components of the committee's suggested framework for informing airborne-PM management (see Figure 2-1 of the present report) that are not adequately addressed in the current draft assessment. Providing less ambiguous recommendations for policy-makers will probably require broader participation in the assessment activities from the full range of PM scientists, including those who study human health, management strategies and their economic effects, and control technology.

The committee recognizes that NARSTO's expertise lies in atmospheric science and that NARSTO has therefore limited its current assessment activities largely to summarizing the state of atmospheric science. At the same time, the committee notes that objectives 5 and 6 of NARSTO's charge for this assessment (see Box 1-1 of the present report) indicate that NARSTO clearly sees the importance of informing and coordinating the research of the atmospheric-science community with that in related fields. Indeed, objective 6 includes a goal to deliver atmospheric-science research products in a way that is useful to the other communities. The committee finds that it is impossible for NARSTO to meet those objectives without enhanced interaction between the atmospheric-science community and related disciplines, such as human health, ecosystem and climate sciences, and economics and other social sciences. If those links are not addressed better, the value of future assessment chapters devoted to related fields will be limited, as is the case for the chapter on health effects in the draft assessment. Without such strengthened cross-disciplinary links, NARSTO runs the risk of reducing the relevance of its science assessments to the larger air-quality management activity.

The committee recommends that, in preparation for future assessments, NARSTO augment its activities to encompass a broader array of researchers in issues pertaining to air quality, such as human health, ecosystems, and economics and other social sciences. NARSTO should establish or increase the prominence of leadership positions in its organization to foster interactions with the other communities. Such activities as standing committees; workshops; agency, industry, and symposium briefings; and similar endeavors should be pursued to engage the other communities more fully, rather than simply consulting a few researchers from each field in preparing the next assessment.

REFERENCES

Abbey, D.E., N. Nishino, W.F. McDonnell, R.J. Burchette, S.F. Knutsen, W.L. Beeson, and J.X. Yang. 1999. Long-term inhalable particulates and other air pollutants related to mortality in nonsmokers. Am. J. Respir. Crit. Care Med. 159(2):373-382.

Avol, E.L., W.J. Gauderman, S.M. Tan, S.J. London, and J.M. Peters. 2001. Respiratory effects of relocating to areas of differing air pollution levels. Am. J. Respir. Crit. Care Med. 164(11):2067-2072.

Brunekreef, B., N.A. Janssen, J. de Hartog, H. Harssema, M. Knape, and P. van Vliet. 1997. Air pollution from truck traffic and lung function in children living near motorways. Epidemiology 8(3):298-303.

CAM (Comisión Ambiental Metropolitana). 2001. Inventario de Emisiones 1998 de la Zona Metropolitana del Valle de México [in Spanish]. [Online]. Available: http://www.sma.df.gob.mx/publicaciones/aire/inventario1998/ver2/inventario1998.htm [September 5, 2002].

CARB (California Air Resources Board). 2001. The 2001 California Almanac of Emissions and Air Quality. Sacramento, CA: California Air Resources Board. 435pp.

Chay, K.Y., and M. Greenstone. 1999. The Impact of Air Pollution on Infant Mortality: Evidence from Geographic Variations in Pollution Shocks Induced by a Recession. National Bureau of Economic Research, Working Paper No. w7442. [Online].Available: http://papers.nber.org/ papers/w7442 [September 5, 2002].

Chow, J.C., J.G. Watson, and S.A. Edgerton, eds. 2002. Thematic issue—Tropospheric aerosols: Science and decisions in an international community. Sci. Total Environ. 287(3):165-286. March 27, 2002.

Cuidad de Mexico. 1999. Towards an Air Quality Programme 2000-2010: For the Mexico City Metropolitan Area. [Online]. Available: http://www.sma.df.gob.mx/publicaciones/aire/prog_cal_aire00_10/2000/ing/capitulo02.pdf [June 13, 2002]

Department of the Environment. 2000. Canadian Environmental Protection Act, 1999. Government Notices. Canada Gazette Part 1, 134(6). (February 5, 2000). [Online]. Available: http://collection.nlc-bnc.ca/100/201/301/canada_gazette-e/part1/2000/g1-13406_e.txt

Dockery, D.W., C.A. Pope 3rd, X. Xu, J.D. Spengler, J.H. Ware, M.E. Fay, B.G. Ferris Jr., and F.E. Speizer. 1993. An association between air pollution and mortality in six U.S. cities. N. Engl. J. Med. 329(24):1753-1759.

Dominici, F., M. Daniels, S.L. Zeger, and J.M. Samet. 2002. Air pollution and mortality: Estimating regional and national dose-response relationships. J. Am. Stat. Assoc. 97(457):100-111.

EPA (U.S. Environmental Protection Agency). 2002. Third External Review Draft of Air Quality Criteria for Particulate Matter (April 2002). EPA/600/P-99/002ac. Office of Research and Development, National Center for Environmental Assessment, Research Triangle Park, NC. [Online]. Available: http://www.epa.gov/ordntrnt/ORD/archives/2002/june/htm/article2.htm [September 6, 2002].

Friedman, M.S., K.E. Powell, L. Hutwagner, L.M. Graham, and W.G. Teague. 2001. Impact of changes in transportation and commuting behaviors during the 1996 Summer Olympic Games in Atlanta on air quality and childhood asthma. JAMA 285(7):897-905.

Hales, J.M. 2002. NARSTO: A Quick Overview. Presentation at the First Meeting of the Committee to Review NARSTO's Scientific Assessment of Airborne Particulate Matter, National Research Council, Washington, DC. April 11, 2002.

Hansen, D.A. 2000. Memo from D. Alan Hansen (EPRI) to Scott Mathias (EPA), August 3, 2000. (as cited in NARSTO Fine Particle Assessment External Review Draft, December 31, 2001).

IPCC (Intergovernmental Panel on Climate Change). 2002. Climate Change 2001: The Scientific Basis, J.T. Houghton, Y. Ding, D.J. Griggs, M. Noguer, P.J. van der Linden, and D. Xiaosu, eds. Cambridge, UK: Cambridge University Press. 944 pp.

Mejia-Velazquez, G.M., and M. Rodriguez-Gallegos. 1997. Characteristics and estimated air pollutant emissions from fuel burning by the industry and vehicles in the Matamoros-Reynosa border region. Environ. Int. 23(5):733-744.

Molina, LT., and M.J. Molina, eds. 2002. Air Quality in the Mexico Megacity: An Integrated Assessment. Dordrecht: Kluwer Academic Publishers.

Mukerjee, S., ed. 2001. US-Mexico Transboundary Air Pollution Studies. The Science of the Total Environment. 276(1-3):1-238.

NARSTO (North American Research Strategy for Tropospheric Ozone). 2000. An Assessment of Tropospheric Ozone Pollution: A North American Perspective. Palo Alto, CA: EPRI.

NRC (National Research Council). 1994. Science and Judgment in Risk Assessment. Washington, DC: National Academy Press.

NRC (National Research Council). 1998. Research Priorities for Airborne Particulate Matter: I. Immediate Priorities and a Long-Range Research Portfolio. Washington, DC: National Academy Press.

NRC (National Research Council). 1999. Research Priorities for Airborne Particulate Matter: II. Evaluating Research Progress and Updating the Portfolio. Washington, DC: National Academy Press.

NRC (National Research Council). 2000. Review of the NARSTO Draft Report: An Assessment of Tropospheric Ozone Pollution – A North American Perspective. Washington, DC: National Academy Press.

NRC (National Research Council). 2001. Research Priorities for Airborne Particulate Matter: III. Early Research Progress. Washington, DC: National Academy Press.

Pope 3rd, C.A. 1989. Respiratory disease associated with community air pollution and a steel mill, Utah Valley. Am. J. Public Health 79(5):623-628.

Pope 3rd, C.A., M.J. Thun, M.M. Namboodiri, D.W. Dockery, J.S. Evans, F.E. Speizer, and C.W. Health. 1995. Particulate air pollution as a predictor of mortality in a prospective study of U.S. adults. Am. J. Respir. Crit. Care Med. 151(3 Pt1):669-674.

Pope 3rd, C.A., R.T. Burnett, M.J. Thun, E.E. Calle, D. Krewski, K. Ito, G.D. Thurston. 2002. Lung cancer, cardiopulmonary mortality, and long-term exposure to fine particulate air pollution. JAMA 287(9):1132-1141.

Samet, J.M., Z.L. Zeger, F. Domenici, F. Curriero, I. Coursac, D.W. Dockery, J. Schwartz, and A. Zanobetti. 2000. The National Morbidity, Mortality, and Air Pollution Study, Part 2. Morbidity and Mortality from Air Pollution in the United States, Research Report 94. Health Effects Institute, Cambridge, MA. June, 2000. [Online]. Available: http://biosun01.biostat.jhsph.edu/~fdominic/research.html. [September 6, 2002].

SCERP (Southwest Center for Environmental Research and Policy). 2002. Studies. [Online]. Available: http://www.scerp.org/ [September 6, 2002].

Schwartz, J., F. Ballester, M. Saez, S. Perez-Hoyos, J. Bellido, K. Cambra, F. Arribas, A. Canada, M.J. Perez-Boillos, and J. Sunyer. 2001. The concentration-response relation between air pollution and daily deaths. Environ. Health Perspect. 109(10):1001-1006.

Shepherd, M. 2002. Presentation to External Review Committee. Presentation at the First Meeting of the Committee to Review NARSTO's Scientific Assessment of Airborne Particulate Matter, National Research Council, Washington, DC. April 11, 2002.

Vet, R., J. Brook, T. Dann, and J. Dion. 2001. The nature of $PM_{2.5}$ mass, composition and precursors. In Precursor Contributions to Ambient Fine Particulate Matter in Canada, J.R. Brook, P.A. Makar, M.D. Moran, M.F. Shepherd, R.J. Vet, T.F. Dunn, and J. Dion. Toronto: Meteorological Service of Canada. (as cited in NARSTO Fine Particle Assessment External Review Draft, December 31, 2001).

Vickery, J. 2002. NARSTO's PM Assessment: Policy Relevance. Presentation at the First Meeting of the Committee to Review NARSTO's Scientific Assessment of Airborne Particulate Matter National Research Council, Washington, DC. April 11, 2002.

Watson, J.G., J.C. Chow, and S.A. Edgerton, eds. 2001. Special issue—Tropospheric aerosols: Science and decisions in an international community—A peer-reviewed selection of papers presented at the NARSTO symposium, Querétaro, Mexico, October 24-26, 2000. J. Air & Waste Manage. Assoc. 51(11):1505-1608. November 2001.

ATTACHMENT A

BIOGRAPHICAL INFORMATION ON THE COMMITTEE TO REVIEW NARSTO'S SCIENTIFIC ASSESSMENT OF AIRBORNE PARTICULATE MATTER

Joe L. Mauderly (Chair) is vice president of the Lovelace Respiratory Research Institute and director of the institute's National Environmental Respiratory Center. He is a former chairman of the Electric Power Research Institute's Air Pollution Health Studies Advisory Committee and a former member of the Health Effects Institute's Research Committee. He was the chair of the Environmental Protection Agency's Clean Air Scientific Advisory Committee and is currently a member of the National Research Council Committee on Research Priorities for Airborne Particulate Matter. Dr. Mauderly received his D.V.M. from Kansas State University and specialized in respiratory physiology and comparative pulmonary responses to inhaled toxicants.

Michael Brauer is professor in the School of Occupational and Environmental Hygiene at the University of British Columbia, Canada. Dr. Brauer's research subjects include particulate air-pollution epidemiology, exposure assessment, exposure to and health effects of biomass combustion, and ozone exposure and health effects. He is a member of several major advisory committees, including the (U.S.-Canada) International Joint Commission Air Quality Advisory Board and the Science Advisory Panel of the US National Urban Air Toxics Research Center. Dr. Brauer received an Sc.D. in environmental health from the Harvard School of Public Health and a B.A. in biochemistry and environmental sciences from the University of California, Berkeley.

Mauricio Hernandez-Avila is director of the Center for Public Health Research of Mexico's National Institute of Public Health. Dr. Hernandez-Avila has consolidated an interinstitutional group that develops research on environmental pollution. His research interests include lead intoxication and air-pollution health effects. He received an M.S. and a Ph.D. in epidemiology from the Harvard School of Public Health and is a graduate of the School of Medicine of the National Autonomous University of Mexico.

Nancy Kete is director of the Climate, Energy and Pollution Program of the World Resources Institute. Her research interests include energy and environmental policy and the use of economic instruments for environmental protection. She was deputy director of the Environmental Protection Agency's Office of Atmospheric Programs. Dr. Kete received a B.S. in geography from Southern Illinois University and a Ph.D. in geography and environmental engineering from the Johns Hopkins University.

Charles E. Kolb is president of Aerodyne Research Inc. His research involves experimental and theoretical studies of the chemistry and physics of trace atmospheric species, chemical kinetics, and combustion chemistry and spectroscopy. Dr. Kolb has served on numerous National Research Council bodies, including the Committee to Assess the North American Research Strategy for Tropospheric Ozone Program, the Committee on Tropospheric Ozone Formation and Measurement, the Committee on Atmospheric Chemistry, and the Board on Atmospheric Sciences and Climate. He received a B.S. in chemistry from the Massachusetts Institute of Technology and an M.S. and a Ph.D. in physical chemistry from Princeton University.

William Leiss is a professor in the School of Policy Studies at Queen's University (Kingston, Ontario) and research chair in risk communication and public policy in the Faculty of Mangement, University of Calgary (Alberta). His research interests include risk communication, risk management, and public policy. Dr. Leiss was president of the Royal Society of Canada in 1999-2001. He received a B.A. in history from Fairleigh Dickinson University, an M.A. in history from Brandeis University, and a Ph.D. in philosophy from the University of California, San Diego, and he has held university appointments in political science, environmental studies, sociology, and communication.

Gerardo Manuel Mejía-Velázquez is professor of environmental and chemical engineering at the Center for Environmental Quality of the Instituto Tecnológico y de Estudios Superiores de Monterrey (ITESM) in Mexico. His research interests include environmental modeling and optimization and parameter estimation. He has worked on joint international research projects related to air-quality modeling and education and is on the advisory board of the Mexican Research and Development Network on Air Quality in Large Cities. He is a member of the Geosciences Comitee of El Consejo Nacional de Ciencia y Tecnología (CONACYT) and of the Technnical Advisory Committee for the National Emission Inventory in Mexico and a board member of the Advanced Technology Environmental Education Center. Dr. Mejía-Velázquez received a B.S. in chemical engineering from Universidad Autónoma de San Luis Potosí, an M.E. in chemical engineering from ITESM, and a Ph.D. in chemical engineering from Texas A&M University.

Luisa T. Molina is an atmospheric chemist in the Department of Earth, Atmospheric and Planetary Sciences of the Massachusetts Institute of Technology and the executive director of the Integrated Program on Urban, Regional, and Global Air Pollution. The central element of the program is an integrated assessment of air pollution in Mexico City. Her research interests include molecular spectroscopy, chemical kinetics, and atmospheric chemistry. She has been involved in particular with the chemistry of stratospheric ozone depletion and urban air pollution. Dr. Molina is a member of the Mexican Research and Development Network on Air Quality in

Large Cities. She received a B.Sc. in chemistry from McGill University and a Ph.D. in chemistry from the University of California, Berkeley.

Lynn M. Russell is assistant professor of chemical engineering at Princeton University and head of the Atmospheric Aerosol Research Group there. Her research interests are in aerosol composition, concentrations, and dynamics in the troposphere. She received her Ph.D. in chemical engineering from the California Institute of Technology. Dr. Russell received her B..S in chemical engineering and A.B. in international relations from Stanford University.

ATTACHMENT B

LINE-BY-LINE COMMENTS

PREFACE

Page iii, line 11: The author must mean "location" of exposure. The "route" of exposure is known (inhalation).

Page iii, footnote 1: It should be "NARSTO is."

EXECUTIVE SUMMARY

Page ES-2, lines 20-21: The wording needs editing. "Hypothesis" does not "interact" with the body to "provide causal explanations."

Page ES-5, line 16: Insert "and" before NO_x.

Page ES-5, line 19: The wording needs editing. The committee does not think you are really talking about "interactions between PM and issues."

Page ES-8, lines 15-24: Should this paragraph also mention Mexico City?

Page ES-9, lines 1-2: Because many readers will read only the executive summary, you should spell out all abbreviations or otherwise explain what they mean.

Page ES-9, lines 8-9: "1 in 6 days or more frequently" would be more explicit than "1 in 6 days or better ."

Page ES-12, line 6: Although much more rare, regulations can also address nonanthropogenic PM emissions, such as pollutants.

Page ES-12, line 16: Why not use "EC" as the preferred term, and not "black carbon"?

Page ES-12, lines 22-23: Evaporative emissions (from vehicles, paints, solvents, and so on) are also substantial sources of organic carbon.

Page ES-12, lines 33-34: It is stated that transport of Asian PM to North America happens a few times per decade. How often is African PM transported to North America?

Page ES-15, bullet 1: These tools can be used to work on solutions, but they are more typically used for detecting problems.

Page ES-15, bullet 2, fifth point: Edit "Reduction in sulfate reduction."

Page ES-18, bullet 1, last line: The last phrase of this sentence is vague.

Page ES-22, line 9: Edit "Other pollution issues other than."

Pages ES-25 and ES-27: These figures are of limited use and are not cited in the text.

Page ES-28, line 32: "is required" should be "was required."

Page ES-29, Table PQ7: It is not clear how "science assessments" and "state-of-knowledge assessments" differ.

Page ES-30, line 30: "Data is available" should be "Data are available."

Page ES-32, line 7: "lease understood" should probably be "least understood."

Page ES-32, line 20: "between pollutants" should be "among pollutants."

Page ES-33, line 17: "data is" should be "data are."

Page ES-34, line 15: Earlier, you describe "ultrafines" as 20 nm, but here you seem to be describing them as 1 nm.

Page ES-34, line 31: "data is" should be "data are."

Page ES-35, line 22: Are you certain that you mean "prospective" here?

Page ES-35, line 25: It should read "studies including data."

Page ES-36, lines 11-12: How about vapors? There is increasing evidence of the importance of organic vapors (semivolatile organic compounds), and they are not generally included in lists of "gases."

Page ES-37, line 24: "characteristics in" should be "characteristics of."

Page ES-38, line 22: Off-road mobile sources should be mentioned; they are not generally included with "transportation."

Page ES-39, line 22: What does "open" sources mean?

Page ES-39, line 27: "insure" should be "ensure."

Page ES-41, line 26: The committee believes that it should read "20% of the organic mass of particles" rather than "20% of organics in particulate matter." Because we cannot speciate it, we do not know what portion of the organic species we can measure. We know that only we cannot speciate more of the organic mass.

Page ES-41, line 31: It should read "operate than filter."

Page ES-43, line 4: It should read "and have been well defined."

Page ES-45, line 33: What are the units for "500"? Miles? Kilometers?

Page ES-47, lines 9 and 34: "CTM" is defined differently in these two lines.

Page ES-47, line 21: It should be "If an emissions tracer...."

Page ES-47, line 27: It should be "...program have cost...."

Page ES-49, line 25: These measures are not "surrogates for PM health effects," nor is it clear how they could be.

Page ES-51, lines 18 and 22: There is nothing wrong with the term "broaching" itself, but this is the first such usage the committee has seen. One usually speaks in terms of "exceeding" standards.

Page ES-52, line 30: Does one ordinarily think of models in terms of having "skills"?

Page ES-53, line 22: It should read "programs comparable with that of the United States."

CHAPTER 1

Page 1-3, Box 1: Although emissions of semivolatile gases contribute to PM, it seems incongruous to include them as "primary PM" emissions. They are not PM when emitted.

Page 1-3, line 4: Elemental carbon is called "black carbon" elsewhere. The terminology should be consistent.

Page 1-3, lines 4-6: Although it is true that the bulk of coarse PM mass comprises crustal materials, it is worth noting that it also contains PM that is not of crustal origin, such as road-surface wear particles, tire and brake wear particles, and vegetation detritus. Indeed, much of urban "road dust" is not crustal in origin.

Page 1-4, Box 2, paragraph 1: What are "black organic liquids"? Is this term explained elsewhere? If it is commonly used, the committee has never seen it.

Page 1-4, Box 2, last paragraph: The issue of adjusting the organic carbon value on the basis of "molecular form" is not presented clearly here. It is an important issue if one wants to understand the present state of science in this field. A reader who does not already understand this issue could not understand it from this description.

Page 1-5, line 9: The statement here is true, but ultrafine PM also results from combustion that is not anthropogenic and from noncombustion sources. Ultrafine PM is not well understood by the health-research community and is even less understood by the lay public. It is worth clarifying the nature and sources of ultrafine PM in this assessment.

Page 1-7, Table 1.2: Should include forest fires and biomass burning in the general source types. They are discussed on lines 11-18 of page 1-7.

Page 1-8, line 10: Delete "daily."

Page 1-8, line 10: The daily mortality and morbidity are important, but long-term effects on disease and survival are also important. Indeed, many now believe that the long-term effects of PM constitute a more serious public-health burden than the short-term effects.

Page 1-10, line 1: Change "equivocally" to "unequivocally."

Page 1-11, Table 1.3, second box down in right column: Many particles are nonspherical, but many are spherical. Third box down in right column: It is true that most ultrafine PM is "transient in character"; i.e., they don't last long in that mode. That's important in considering the relationship between health effects and proximity to fresh combustion emissions. However, a more important issue for health is whether ultrafine PM exists where people breathe, not how old the PM is. Most people live and work in areas where there is always considerable ultrafine PM. Last box down in right column: Why say "in most places"? Is there any common site where there are not variable mixtures of many pollutants? This wording denies the true predominance of the issue.

Page 1-15, line 7: Add "Mexico."

Page 1-17, footnote 5: There is a more specific reference for this document (it has an EPA document number and date).

Page 1-20, Figure 1.7: This figure needs a bit of attention to make it clear (whether or not it is represented exactly as originally published). Not all PM types are not found in the scheme. That is alluded to in the text but not in the figure legend. The tier labeled "BC and OC" shows the OC hierarchy, but not BC. The OC hierarchy is unclear. For example, both primary and secondary OC leads to OC PM, and the "anthropogenic vs biogenic" issue starts at the top, not just at the bottom of the hierarchy. The concepts that this figure is intended to portray are important, but they are not presented well by the figure. The authors would do better to develop a new figure.

Page 1-20, Figure 1.8: The figure legend is misleading. The figure does not illustrate only processes relevant to PM "formation." For example, deposition is not a "formation" issue.

CHAPTER 2

Page 2-1, lines 23-24: Language is awkward.

Page 2-1, lines 26-28: Sentence implies that ammonium nitrate homogeneously nucleates rather than forming heterogeneously on pre-existing particles; this has not been demonstrated.

Page 2-3, lines 6-7: "water leaves the particle and returns to the vapor phase."

Page 2-7, line 9: Change "because of" to "against."

Page 2-8, line 10: At the end of the line, add "However, it has been shown that black carbon and organic aerosol react with OH and other oxidants, making them hydrophilic; hence, the particles absorb water more efficiently as they age (see Bertram, A.K, A.V. Ivanov, M. Hunter, L.T. Molina, and M. J. Molina. The reaction probability of OH on organic surfaces of tropospheric interest. *J. Phys. Chem.*, 105, 9415-9421, 2001)."

Page 2-12, line 12: "lead to the reduction of particulate nitrate and returns nitrogen oxides to the gas phase."

Page 2-16, line 1: Give references to the studies referred to in this paragraph.

ATTACHMENT B   65

Page 2-16, lines 9-10: Phrasing is awkward phrasing. How about "From Precursor Emissions to Aerosol Component Concentration"?

Page 2-16, lines 26-28: Language is awkward.

Page 2-18, lines 24-25: Refer to table data (in blue). Are the tables going to be in color?

Page 2-19, lines 17-19: There is a minor logical problem. Sulfate reduction does not directly result in "freed" ammonium; without the sulfate, ammonia would not have been in the ammonium form. It does leave more unscavenged gas-phase ammonia to react with nitric acid, as the next sentence states.

Page 2-21, Figure 2.10: This figure is confusing; perhaps it could be improved, although the topic is complicated.

Page 2-24, line 30: There is a minor logical problem in "their activation and the resulting droplet's subsequent lifetime."

Page 2-26, line 9: Change "from Asia and" to "Asia to North America and." Cross check and refer to Appendix D.

Page 2-27, line 3: "to HNO_3 which reacts with available ammonia."

Pages 2-26 and 2-29: Add subheadings to separate the PM-ozone discussion from the PM-haze discussion, if it is desirable to retain the PM to haze discussion.

Page 2-30, line 15: Update the reference "(IPCC, 1995)" to "(IPCC, 2002)" and the corresponding material as needed.

CHAPTER 3

Page 3-1, line 5: There is an extra comma after "PM."

Page 3-2, line 4: Table 3.1, title "10 _m" should be "10 µm."

Page 3-4, line 6: "green house" should be "greenhouse."

Page 3-4, line 9: Define "CTMs" because it is used here for the first time.

Page 3-6, line 8: The reference cited under footnote 1 of Table 3.2 should not have "e.g.," if the reference refers to only one special study.

Page 3-7, line 17: "Molinas et al. (2001)" should be "Molina et al. (2000)." (Perhaps the latest reference, Molina and Molina [2002], should be used.)

Page 3-7, line 24: Delete the gap between "PM" and subscript "10" and between "PM" and subscript "2.5."

Page 3-8, line 2: Change "are not necessarily relevant to the vapors that are believed to produce secondary organic carbon" to "are unlikely precursors of secondary organic carbon particulate matter."

Page 3-9, Table 3.3: Check the numbers listed under "Open Sources" and "Fugitive Dust"; footnote 2 indicates that fugitive dust is part of open sources, but the values for fugitive dust under "Primary PM_{10}" and "Primary $PM_{2.5}$" for the United States and Canada are larger than the values for open sources. The last column, fugitive dust is listed as contributing "4322" to NH_3 for the United States? Footnote 1 indicates that bold numbers are emphasized for large contributions; but in the last column, "88" us boldface but not "270."

Page 3-10, line 11: What is "VOC_{part}"?

Page 3-11, Table 3.4: The unit for Max PM10 should be $\mu g/m^3$. Specify the year when the data were reported, such as the population of Mexico City and the peak O_3 of 300 ppb. Data on Mexico are suspicious. For example, the GDP is listed as US$ "2400" (what year?) but the value is US$ 7750 in 2000. The committee suggests referring to Table 2.1 (page 22) of Molina and Molina (2002), which also includes other statistics. Footnote 4: What does "nominal value" mean? The total area of the Mexico City metropolitan area is 5300 km^2, and of the urbanized area 1500 km^2. The committee suggests referring to Table 3.5 (page 74) of Molina and Molina (2002).

Page 3-13, Table 3.5, last column: PM_{10} includes $PM_{2.5}$, so the value for Los Angeles "on road" $PM_{2.5}$ emissions of 9200 is inconsistent with the value of 7800 for PM_{10}. Footnote 3: Check spelling for CAM; it should be "Comisión Ambiental Metropolitana"?

Page 3-14, Table 3.6: The value of the on-road contribution to NOx for Mexico City should be "165,800," not "65,800." Footnote 3: Check spelling for CAM; should it be "Comisión Ambiental Metropolitana"?

Page 3-15, line 2: Assuming that the values listed in Table 3.5 were correct, the primary PM_{10} emission for Mexico City is about one-seventh to one-sixth of Toronto's or Los Angeles's emission, so, the phrase "perhaps amounting to only one-third to one-fourth of Toronto's PM_{10} emissions, and even less than Los Angeles" is incorrect.

ATTACHMENT B   67

Page 3-15, line 5: The phrase "along with transportation sources" is inconsistent with the values listed in Table 3.5 and on line 4 of page 16; that is, the contributions from "open sources" are much larger than from "transportation sources."

Page 3-15, lines 11 and 12: According to Table 3.5, the ratio of $PM_{2.5}$ to PM_{10} for Atlanta should be 24%, not 34%. Therefore it is not "about the same as in Los Angeles (36%)." However, note that there is a question about the $PM_{2.5}$ value for on-road $PM_{2.5}$ emissions for Los Angeles as commented on above.

Page 3-16, line 2: The assumption that "all of the PM observed in the cities is local in origin" is debatable and inconsistent with line 13.

Page 3-16, line 4: "0.1" should be "0.16" or "0.2" if rounded off to one significant figure. However, see line 5, which uses two significant figures: "0.34 to 0.55."

Page 3-16, line 5: "0.5" should be "0.8."

Page 3-16, line 12: Rephrase the first sentence from "Another dimension reflected in the emission inventories is the geographical distribution of pollutants." to "The geographical distribution of pollutants results from the emissions and dispersions of pollutants."

Page 3-25, line 3: "ands" should be "and."

Page 3-25, line 29: "2001" should be "2000." The committee suggests referring to discussions in Section 4.1 starting on page 164, citing particularly page 167, of Molina and Molina (2002).

Page 3-26, line 14, "(" should go to the next line.

Page 3-29, line 20: Clarify "resolution of definitions." Does it mean resolution of differences, resolution of uncertainties, or resolution of definitions? If the latter, which definitions?

Page 3-31, lines 6-9: It should be noted that Mexico is using modified MOBILE 5 or MOBILE5a.3MCMA for VOC, NO_x, and CO but still uses the U.S. method for PM_{10}. The committee suggests referring to Section 6.3 on page 201 of Molina and Molina (2002) and page 24 of CAM (2001).

Page 3-33, line 22: The source profiles developed in the middle 1970s in Los Angeles should be reviewed; there is a chance that they have seriously shifted in as much as major changes in emission control of vehicles have taken place since that time.

Page 3-38, line 8: The committee suggests checking with the Mexican authorities about Mexican laws concerning disclosure of industrial pollution reported to the government.

Page 3-39, line 18: There are extra spaces before the commas.

Page 3-39, lines 27-28: Concerning the application of the MOBILE model into a GIS positioning system, the committee suggests checking the NRC report (National Research Council. 2000. Modeling Mobile Sources, National Academy Press, page 48) on the limitations of the MOBILE model.

Page 3-40, line 19: Fine particulate matter does not entirely follow population centers except in the broadest sense. For clarity, add the word "sources" after "$PM_{2.5}$."

Page 3-41, line 20: What is meant by "traditional"? Man-made or natural? The committee suggests deleting the phrase "along with more traditional sources."

CHAPTER 4

Page 4-7: The descriptions of the different types of size-selective inlets would benefit from simple diagrams. The descriptions are fine for those who are already familiar with the inlets; for those who are less familiar, the text descriptions do not adequately describe how the inlets work.

Page 4-7, line 34: If these are used in both urban and nonurban areas, what areas are left?

Page 4-9, line 28: Fix "may the altered."

Page 4-10: Recommendation 2.5 should be recommendation 1.3. Recommendation 1.3 (page 4-25) should be 1.6.

Page 4-12, line 13: The pressure-drop continuous mass-measurement device is mentioned but not described; some description should be added.

Page 4-28, lines 14-15: What are the problems with current approaches to calibration for O_3, NO and NO_2? These are not indicated in Table 4.4, nor are they described in the text.

Pages 4-15 and 4-31: Personal monitors can achieve much better accuracy than a factor of 2 for major ionic species relative to ambient measurements (such as of sulfate); in fact, agreement is usually better than for PM mass. There are commonly used photometers that have been used for real-time personal monitoring and these should be mentioned. Also, much of the early personal-monitoring work was done with portable piezobalances.

Page 4-16, lines 14-16: The statement that single-particle measurement methods are "available only as research tools" is outdated and needs to be deleted or changed. Two of the instruments listed in Table B.4 are now commercially available: the ATOFMS from TSI and the AMS from Aerodyne Research, Inc.

ATTACHMENT B   69

Page 4-27, Table 4.3: Should acknowledge the 20% of the organic (speciated) that is known.

CHAPTER 5

Page 5-2, line 13: Fix "Topography, through its influence circulation, strongly affects."

Page 5-2, line 20: "Prevailing air mass patterns (see Fig. 5.1) make it possible."

Page 5-4, Table 5.1; "Comment" 2: "Scale of concern for climate change." Also, spell out or define "PSD"; it is not in the glossary.

Page 5-5, line 19, and page 5-6, line 22: Effects on "Central America and the southeastern United States or U.S. Gulf Coast" are cited twice. This is a trinational report, so we should probably acknowledge that Mexico lies between those two regions and is also affected.

Page 5-7, Section 5.3: This section identifies eight "different" North American regions. Chapter 10, which is the other major chapter with strong policy relevance, identifies nine, separating Los Angeles and the San Joaquin Valley.

Page 5-8, lines 5-6: The "Valle de Mexico" basin is *NOT* 1300 km^2. The "Basin of Mexico" includes most of the Federal District, part of the state of Mexico, and southeast parts of the state of Tlaxcala and southern parts of the state of Hidalgo (Tizayuca and other municipalities). It also includes the main topographic features (volcanoes, mountains, and so on). All this area is about 9560 km^2. However, the "Valle de Mexico" (Valley of Mexico or Mexico City metropolitan area) is the nearly flat floor of this basin that occupies a total area of about 5300 km^2, of which about 1500 km^2 is urbanized.

Page 5-11, Fig 5.4: Data on El Paso/Juarez and labeled as just "El Paso/".

Page 5-12, lines 12-13: It is not clear that the cited Figure 5.3 supports the point made, because paired urban-nonurban monitoring sites are not identified in the figure.

Page 5-13, Figure 5.5: This has poor contrast and is hard to read.

Page 5-14, Figure 5.6: There are far too many significant figures on the axes.

Page 5-16, Figure 5.7: Data on Mexico City are not included (although they are available and presented in Figure 5.12).

Page 5-20, Figure 5.9: Labels are scrambled.

Page 5-21, Figure 5.10: This is too small.

Page 5-21, line 5: The phrase "but neither are these rare events" is repeated. The committee recommends that these statements be quantified rather than saying only that they are "not rare." For example, do the events occur 10% of the time or less?

Page 5-23, Figure 5.11: The legend is very unclear, as is the meaning of the graphs. The red circle symbol seems to be entirely missing from the graph. The figure is good, but there are faded overprints, and locating arrows are needed for the bottom three plots.

Page 5-26, Figure 5.12: The outline of Mexico is omitted from the map

Page 5-27, Figure 5.13: Change SO_4 to either sulfate or SO_4^{2-}; be consistent with rest of report.

Page 5-32, Fig 5.15: This is generally illegible. Not enough is legible for reading this graph or even to know what pollutants were measured.

Page 5-34, Figure 5.16: This is too small, and site labels are missing.

Page 5-35, Figure 5.17: This is too small, and site labels are missing.

Page 5-40, Figure 5.18: This is too small.

Page 5-43, line 18 and elsewhere: Hidy 19XX is referenced. Add year of publication.

Page 5-44, Figure 5.20: The scale of the graphs makes it most difficult to evaluate any trends or changes.

Page 5-46, lines 1-5: Needs a reference for the stated trends.

Page 5-46, line 7: How is "improvement in PM exposure" defined and is it true that it has improved? Even though concentrations have gone down, there are larger populations in urban areas, so it might be possible that exposure (total rather than per-person) has gone up. It is also important to mention that in Mexico PM exposure is now worse than it was 50 years ago.

Page 5-47: There is inconsistency between the statement that PM in California "is dominated by ammonium nitrate and carbon compounds" and the earlier statement that the Grand Canyon gets a substantial fraction of its sulfate from California. They cannot both be true unless there is vast SO_2 conversion between California and the Grand Canyon.

Pages 5-50 and 5-51, Figures 5.22 and 5.23: Axis labels are skewed.

Page 5-56, line 9: Word choice "principal specific objectives."

ATTACHMENT B   71

Page 5-56, lines 14-15: "variety *of* PM carbon."

CHAPTER 6

Page 6-2, line 1: "at long term to evaluate."

Page 6-3, line 2: Replace "of insights that are possible are listed" with "of possible insights are listed."

Page 6-3: In the examples on page 6-3, it would be convenient to comment on how the studies helped to answer policy questions.

Page 6-4: Improve quality of figure. Change legend from "MER Neighborhood-Scle" to "MER Neighborhood-Scale."

Page 6-6, legend: Omit "just from, and were not."

Page 6-8, lines 5-22: What can be said about reliability of meteorology data and their influence on the results?

Page 6-9: The description of Figure 6.4 does not refer to part 3.

Page 6-9, Section 6-2: It would be interesting to have an evaluation of the use of the different methods to support decision-making.

Page 6-10, lines 1 and 2: These lines are repeated.

Page 6-10, line 1: "however" is redundant; the cost is the same.

Page 6-10, lines 1-2: "Costs for receptor-oriented studies range from ~$US 50K for a small community with an established air quality network to ~$US 10M for a large region with a": do you mean source rather than receptor?

Page 6-10, lines 11-12: "are seldom available in retrospect and thus" is awkward and should be reworded.

Page 6-11, table legend: Omit "JEFF state that all methods."

Page 6-14, lines 1-2: Omit "It is the responsibility of the modeler, not the CMB model, to evaluate" and change the remaining part to "The CMB uncertainties and performance measures and such other data as emission inventories should be evaluated."

Page 6-18, paragraph 2, line 4: "depending upon size."

Page 6-19, line 14: Change "than the 10 μm" to "than the original 10 μm emitted size."

Page 6-20, paragraph 3, line 1: In addition to soil and dust....

Page 6-20, paragraph 4, line 1: Figure 6.5b shows a profile of a coal-fired power station. The profile contains many of the same....

Page 6-22, legend: Change "wood burning from typical of Denver" to "wood burning typical of Denver."

Page 6-32, line 3: "before, during and after the peak gives."

Page 6-34, lines 15-16: Change "improved catalyst performance and these were linked, using an aerosol evolution approach, to secondary nitrate and organic aerosol in the ambient PM2.5." to "improved catalyst performance. These emissions were linked by using an aerosol evolution approach to secondary nitrate and organic aerosol in the ambient PM2.5."

Page 6-34, line 19: Omit "thus,".

Page 6-34, line 25: Omit "Even still,".

Page 6-36, legend: Omit ", given the range of sources assumed to be important,".

Page 6-37, in table: Change "effected" to "affected."

Page 6-37, Box: "Responds to decreases" is awkward and unclear; it should be reworded.

Page 6-38, line 3: "bold green box" is missing.

Page 6-40, line 1: Change "markers was found" to "markers were found."

Page 6-42, Figure 6.10: Specify the type of "particle concentration"; is it $PM_{2.5}$, PM_{10}, or something else? It is unclear what is meant by "Fractional change of particle plus gas ammonia concentrations." Is this essentially total ammonia species, that is, particulate ammonium (and ammonia hydrate) + vapor-phase ammonia? If so, please clarify.

This chapter uses "SO4," "SO_4," and "sulfate" to describe the particulate form of S(VI). The chapter (and the entire report) should be consistent in the use of terminology even though the field as a whole is not. In addition, if the chemical formula approach is used, consistently use the proper form indicating the anionic state, "SO_4^{2-}" rather than "SO4" and "SO_4".

Page 6-45, Table 6.2: Define "ROME."

Page 6-45, Table 6.2, second column, 8th box: Unhyphenate "dis-persion."

ATTACHMENT B   73

Page 6-45, Table 6.2, third column, 2nd box: A space is needed before "at receptors."

Page 6-46, Figure 6.11: The legend is blank, and symbols are not defined.

Page 6-47, Section 6.5: Is there an example of Mexico?

Page 6-48: Change paragraph 1 of Section 6.5.2 to read "Observational methods and receptor models are necessary but not sufficient for identifying contributing source types and quantifying their contributions to PM. These methods are heavily dependent on the measurements available, and quantitative apportionment is seldom accurate when applied to measurements that have not been taken for this specific purpose. Receptor methods have strengths and weaknesses. Multiple methods help to characterize uncertainties and concurrence of results. Conclusions. . . ."

Page 6-49: Paragraph 3 seems to be a different section, like a suggested method for source apportionment or attribution. In the same paragraph, line 2, "follow" is repeated; change "following" to "next."

Page 6-49, lines 3-4, last sentence: Change to "Agreement between the receptor and source models increases confidence and represents the optimum 'weight of evidence' outcome."

Page 6-49, lines 9-10: Change "provides clearer focus" to "provides a focus."

Page 6-49, lines 20-21: Change "identify which factors . . . are most important and/or most uncertain" to "identify the factors . . . that are most important or most uncertain."

Page 6-49, line 24: Change "Receptor models need to know which" to "Receptor models require information on."

Page 6-49, lines 34-35: Sentence is awkward. Change to ". . . but these profiles do not necessarily represent those which affect an area where they were not measured."

Page 6-50, point 8: Change "since no model, source or receptor, is" to "Because no source or receptor model is."

Page 6-50, point 8: Awkward word choice. Change to "results must be independently challenged."

Page 6-50, Section 6.6: Sentence starting "Receptor models are useful" is not properly structured and lacks parallelism.

Page 6-50, Section 6.6, line 5: Change "are not usually sufficient, by themselves, to characterize" to "are not usually sufficient to characterize."

Page 6-50, Section 6.6, lines 7-8: Change "The approach to deal with the uncertainties and to obtain the best understanding" to "Reducing uncertainties and improving our understanding."

Page 6-50, Section 6.6, line 11: Omit ", which tend to be empirical,".

Page 6-50, line 13: Change "of the measurements, ambient and source," to "of ambient and source measurements."

Page 6-50, line 13: Change "as well as" to "and."

Page 6-51, paragraph 3, line 9: "Conclusions drawn from this 'weight of evidence' approach will be more defensible for decision-making, even when there are disagreements, and agreement between approaches clearly increases confidence" is awkward.

Page 6-52, Section 6.6.1.1, line 1: Omit "This can be done reasonably well."

Page 6-52, Section 6.6.1.1, line 3: Omit ", rather than absolute,".

Page 6-52, Section 6.6.1.3, line 2: Replace "inorganic secondary material" with "inorganic primary material"; the inorganic secondary components have been named earlier (sulfate, nitrate, and ammonium).

Page 6-53, Section 6.6.1.4, line 4: Replace "expected to more or less" with "expected to be more or less."

Page 6-53, Section 6.6.1.4, line 6: Omit "obviously,".

Page 6-53, Section 6.6.1.4, line 8: Omit "upon."

CHAPTER 7

Page 7-4, Figure 7.1: This figure is not explained in the text or in the legends. Labels are missing from the arrows.

Page 7-5, line 6: "Since" should be "Because." The same change should be made on page 7-8, line 23; page 7-11, line 14; page 7-15, lines 9 and 15; page 7-19, line 21; page 7-20, line 21; page 7-21, line 27; page 7-22, line 21; page 7-27, line 14; page 7-42, lines 21 and 24.

Page 7-5, line 24: Rather than "kilometers traveled," such activity ought to refer to the "kilometer-tons" traveled, that is, the amount of work done. Pollution per vehicle would be put in much better management perspective if pollution were dealt with in terms of the unit of work vehicles do. With regard to societal benefit, a large truck probably emits less pollution per ton of load moved than an SUV.

Page 7-8: Some mention should be made of on-line models, i.e., those models in which the chemistry, emissions, deposition, and other air quality algorithms are embedded in the meteorology code and thus can make full use of the meteorology at the highest spatial and temporal resolution. Although such models are not yet widely used, they look to be the future of CTM development and therefore warrant at least a brief discussion.

Page 7-8, line 29: Put a comma after "scales."

Page 7-10, lines 3-5: To many, if not most, readers of this document, the information here would be incomprehensible (for example, "order-1-5 TKE scheme . . . ").

Page 7-10, line 4: Explain briefly what is the TKE scheme and the order-1 K Theory.

Page 7-10, line 8: The concept might be clearer for most readers if "grid size" were used instead of "grid discretization." This pertains to later use as well.

Page 7-12, lines 23-24: Not only is this not realistic for fresh emissions, but it is not realistic for any environmental collection of PM. There is always PM that is not "internally mixed."

Page 7-13, line 20: Add "structures" to "soil and vegetation." A lot of PM is also deposited on human-made structures of nearly all types.

Page 7-13, lines 23-25: The statement is true, but it is also true that under given circumstances, a model could overestimate or underestimate emissions.

Page 7-14, line 16: The footnote isn't helpful here. If specific examples are to be given in a footnote, they need to be more understandable to the average reader than the present one.

Page 7-15, line 13: Episodic simulations are not only important for ozone because of the time criteria for the standards; they are also important because many air-quality management districts would like to be able to give advance public-health warnings.

Page 7-15, line 18: What is a "chemical lateral boundary"?

Page 7-19, Section 7.4: Explain how the questions were formulated. This may be explained if at the beginning of the chapter questions 1-4 are answered.

Pages 7-20 and 7-21, Sections 7.4.2 and 7.4.3: The answer is "yes," but then it is explained how the answer may be "not". This should be consistent. If "yes," state what is available and what is needed.

Page 7-23, line 9: The metrics that are mentioned here are not, nor could they be considered, "metrics for PM adverse health effects." The committee understands the "surrogate metrics" concept being portrayed, but understanding of the relationship between health effects and PM measures is so uncertain (albeit there is confidence that there is one) that one could not use these as surrogate metrics for health effects.

Page 7-24, line 26: To provide a correct definition, it should be "diameter of a spherical particle."

Page 7-30, Table 7.1: In the next to last line of the title, "between" should be "among."

Page 7-32, line 16: PM dose also depends on the inspiratory rate (presumably, "mode of breathing" means oral vs nasal) and anatomic differences among the airways of individuals, in addition to particle size and hygroscopicity.

Page 7-32, lines 31-32: The concentrations in motor vehicles will also depend on the number, type, and emission rates of the other vehicles on the road that contribute to the air parcel encountered by the vehicle of concern and to wind direction and speed.

Page 7-44, lines 9-10: The "4-D" concept should be explained; not all readers will understand what is meant.

Page 7-44, line 26: "performance is best for sulfate, and typically more uncertain for nitrate and ammonium."

Page 7-47, line 21: "Results" should be "Result."

Page 7-48, line 22: It would be useful to mention the record for accuracy here. If the models have been used to forecast ozone levels, what is their record of accuracy?

CHAPTER 8

Page 8-1, paragraph 3, line 5: "Role" should be "Roles."

Page 8-2, Table 8.1: The list of chronic effects in the footnote conflicts with the term "acute" in the title.

Page 8-3, line 1: Do hospital discharges include deaths in the hospital?

Page 8-3, paragraph 1: What about lung cancer? There is a statistically significant link between PM and lung cancer in long-term studies, and this issue has recently drawn more attention.

Page 8-3, Table 8.2: What about the Northeast region? It seems odd that this (most populous) region is left out.

Page 8-3, Table 8-2: What is the relevance of presenting regional incidence data only for the United States? Is there supposed to be some inference that these differences are due to PM? In this context, the table might be misleading.

Page 8-4, paragraph 3: What about outdoor exposures to PM originating indoors? Is that "ambient" or "non-ambient"? Although one would not expect indoor-origin PM to be a major source, it is undeniably one of the sources (and one that is virtually never mentioned).

Page 8-4, Section 8.3, line 1: This sentence should be clarified; it is not restricted to personal exposure "in each microenvironment" but to personal exposure itself. "Non-ambient" and "ambient" PM should be described (as in the criteria document and primary literature) as PM of ambient and nonambient *ORIGIN*. The studies that have evaluated the relative contribution of ambient and non-ambient sources should be summarized (about 50% of total exposure is to PM of ambient origin). Furthermore, the importance (or lack thereof) of the distinction between PM of ambient and nonambient origin with respect to epidemiology should be described; for example, that there are substantial exposures to PM of ambient origin does not mean that ambient-PM epidemiology is invalid.

Page 8-4, last paragraph: Do these statistics refer only to the United States?

Page 8-5, paragraph 1, line 2: "Since" should be "Because."

Page 8-5, paragraph 1, line 5: "This data" should be "These data."

Page 8-5, paragraph 3, line 1: "Between" should be "Among."

Page 8-6, paragraph 3, line 1: What does "relatively" mean here? Relative to what?

Page 8-7, Figure 8.1: If the figures from the Environmental Protection Agency PM CD are taken from some other original source, the original source should be cited, not the CD.

Page 8-8, paragraph 1, last line: See Sarnat at al. (2001), which evaluated the relationship (and its implications for epidemiology) between exposures to multiple gaseous pollutants and PM components: Sarnat, J.A., Schwartz, J. Catalano, P.J., Suh, H.H. 2001. Gaseous pollutants in particulate matter epidemiology: confounders or surrogates? *Environ. Health Perspect.* 109: 1053-1061.

Dosimetry section: Physiologic factors (oral or nasal breathing, disease status, airway size, and so on) are of equal importance as aerodynamic properties of particles in determining deposition.

Page 8-8, paragraph 2, lines 1-2: According to the definition presented here, "dose delivered to" includes consideration of clearance, as well as deposition. Both deposition and clearance act to determine the concentration in the tissue at any given time.

Page 8-8, paragraph 3, line 1: "Airborne inhaled" is redundant. (unless one is drowning).

Page 8-8, paragraph 4, lines 3-5: The statement about growth of hygroscopic PM is correct most of the time but not all the time. If the inhaled air is already saturated at body temperature, hygroscopic PM presumably would not grow after inhalation.

Page 8-8, paragraph 5, line 1: When referring to "large" or "small" particles, give an example as a reference. Otherwise the reader doesn't know what is meant. Even among knowledgeable researchers, these terms have no standardized meaning.

Page 8-8, paragraph 5, line 2: Is "respired" being distinguished from "inhaled"? If so, what is the point?

Page 8-8, last paragraph: A distinction should be made between "respirable" and "inhalable," the two seem to be used interchangeably here, but by convention the do *not* refer to the same thing (occupational exposure limits). The descriptions of the different types of epidemiologic study should focus on the relevant designs in air-pollution epidemiology (or for multifactorial diseases in general). For example, one might not characterize the case-control study as being of highest inferential quality for air-pollution epidemiology.

Page 8-9, Figure 8.2: Cite the original source of the figure.

Page 8-9, paragraph 1: Overall, the discussion of particle size vs deposition is sloppy and needs reworking. The largest particle that can be inhaled is certainly not 10 μm. A 20-μm particle can reach the alveoli—with a low probability. The largest particles that could be inhaled would be of concern for nasal sites. Nasal passages are technically "airways," but they would not ordinarily be called that without the modifier "nasal," and most of this document's audience would not understand that the nose was meant. It would also be useful to mention nose vs mouth breathing. The statement that alveolar deposition is of primary concern for small particles is a half-truth; the highest deposition of the smallest particles is in the nose.

Page 8-10, Figure 8.3: A figure showing deposition curves down to PM below 0.1 μm would be more informative. Overall, this is not a good choice of figures from among the many that have been published.

Page 8-10, paragraph 1: Although aerodynamic size ceases to be important after deposition has occurred, physical size remains important. Physical size of poorly soluble PM can affect its disposition. For example, some researchers are now confident that PM in the 50-nm or smaller range (perhaps even up to 100 nm) either will not be picked up by macrophages or will be scavenged at very low efficiency.

Page 8-10, paragraph 1, line 3: It is not the "total mass" of PM in the air that is important, but the mass concentration, which is different. The committee has not seen an estimate of the total mass of PM in the air, although it would be interesting for conversational purposes.

Page 8-11, paragraph 1: A seventh mechanism is direct penetration of PM into and through cells, which is known to occur.

Page 8-11, paragraph 2, line 1: PM-derived material, whether soluble or nonsoluble, can also be transferred via lymph and may or may not reach the blood via that pathway.

Page 8-11, paragraph 3, line 8: Figure 8.3 doesn't indicate the "smallest airways," as suggested by the text; it aggregates all conducting airways.

Page 8-11, paragraph 3, line 12: "Since" should be "Because."

Page 8-11, paragraph 4, line 1: It should read "particles inhaled," not "inhaled particles."

Page 8-12, paragraph 2, line 9: "Subpopulations" might be a better term for these subdivisions of the general population.

Page 8-13, paragraph 3, line 1: Time-series studies are informative for short-term effects, but other types of studies are more informative for longer-term effects.

Page 8-13, paragraph 4, lines 10-12: It is exposure scientists that (among other things) evaluate the relationship between community estimates and personal exposures. Is this really something in which atmospheric and health scientists need to interact?

Page 8-14: The distinction is not particularly important, but copollutants and meteorologic conditions can be considered both potential confounders and effect modifiers.

What is the relevance of Figures 8.4 and 8.5? Their inclusion may raise more questions than answers.

Page 8-14, paragraph 4: Here, or somewhere in the section on epidemiological evidence, the recently-revealed difficulty with the GAM software should be mentioned. An in-depth discussion is not in order, but rather something that brings the review up to date by making the reader aware of the problem. It would be appropriate to cite the Dominici et al. paper in the *American Journal of Epidemiology* (156(3):1-11, 2002).

Page 8-15, paragraph 3: The information given here is not sufficient for readers to understand Figure 8.6, unless they are already familiar with this type of data. If the figure is worth showing (and it probably is), give it a bit more explanation.

Page 8-19, paragraph 1, line 10: It can be argued that the "challenge" also includes improved understanding of the relationship between source and ambient concentration.

Page 8-20: There have been some initial, provocative attempts to address the issue of the heterogeneity of effect estimates with PM epidemiologicstudies (Janssen et al. APHEA-II), and these should be discussed:

Janssen, N.A., J. Schwartz, A. Zanobetti, H.H. Suh. 2002 Air conditioning and source-specific particles as modifiers of the effect of PM(10) on hospital admissions for heart and lung disease. *Environ. Health Perspect.* 110: 43-49.

Atkinson, R.W., H.R. Anderson, J. Sunyer, J. Ayres, M. Baccini, J.M. Vonk, A. Boumghar, F. Forastiere, B. Forsberg, G. Touloumi, J. Shwartz, and K. Katsouyanni. 2001. Acute effects of particulate air pollution on respiratory admissions: results from APHEA 2 project. Air Pollution and Health: a European Approach. *Am. J. Respir. Crit. Care Med.* 15: 1860-1866.

Page 8-21, Figure 8.7: The quality of the figure needs to be improved. An explanation of the term "lag" and its implications should be given in the text. Many readers might not be familiar with it, and the concept is very important.

Page 8-22, paragraph 2, lines 1-3: There should be a better explanation of the concept of mortality x monitoring days here. The concept is cited several times later and is important. Make certain that all readers understand it well the first time it is introduced.

Page 8-23, line 1, and elsewhere: Statistical power will affect the size of the confidence intervals but should not a priori affect the magnitude or sign of the effect estimates.

Page 8-27, paragraph 1, line 1: Is there *any* ambient environment in which people are not exposed to complex mixtures of pollutants? The wording denies reality by suggesting that mixtures might not always be present.

Page 8-27, paragraph 1, line 7: Like other studies that necessarily relied on measurements of NAAQS pollutants, the NMMAP study only looked at a *few* non-PM pollutants. The statement that it looked at "other" pollutants may be true, but it suggests something much more comprehensive than was actually done. No epidemiologic study to date has included more than a tiny fraction of the pollutants present. This is an important point to make in this chapter.

ATTACHMENT B   81

Page 8-28, last paragraph: If the extended analysis of the AHSMOG cohort, the Peters et al. (1999) study of southern California school children, and the Raizenne et al. (1996) study of 24 Cities in the U.S. and Canada are mentioned, some brief description of their results should be provided.

Page 8-28, paragraph 3, line 1: It is not clear what is meant by "semi-individual."

Page 8-29, paragraph 1, last sentence: State more clearly what "coordinated efforts" are being talked about here. It is of little use to call for more interactions between atmospheric and health scientists if it is not explained what those interactions should be, no example of the envisioned interaction is given, and what could be achieved by such interactions is not stated.

Page 8-29, paragraph 2, last sentence: It should also be mentioned that children have greater ventilation per unit of body size than adults.

Page 8-30, paragraph 3, lines 10-11: First, explain "spline curves"; most of the readers of this document won't know what that term means. Second, perhaps it is meant that the *slopes* increased, rather than that the *curves* increased.

Page 8-30, paragraph 3, line 14: Explain "Akaike's Information Criterion" and give a reference. It is not clear that this term or concept is important to the chapter. If something as arcane as this approach is necessary for the reader to understand, it bears explaining.

Page 8-32, paragraph 1, last sentence: The Southern California Children's Health Study is important for looking at lung-growth effects. This effect and the fundamental findings to date should be discussed.

CHAPTER 9

Page 9-3, line 22-23: This is true, but smokestack plumes can cover broad areas. From low altitudes, one can see a brown cloud layer over a large portion of northwestern New Mexico on many mornings. The wording suggests that non-PM smokestack emissions are only a local problem.

Page 9-8, line 2: It should be "Figure 9.4," not "Figure 2.4."

Page 9-13, Figure 9.5: The source of the figure, or the information in the figure is not given in the figure legend.

Page 9-16, line 16: Define "ASOS." Visibility information is also given by AWOS (automated weather observation system [or service, I don't remember which]) at airports. Why not cite that as well.

Page 9-16, recommendation box, last sentence: The term "better" isn't very descriptive. Say "shorter," "longer," "more frequently," or whatever is meant.

Page 9-31, lines 19-21: The issue of the humidity effect on the relationship between PM concentration and visibility could be presented more clearly. If PM is measured under standard humidity conditions—that is, if filters are dried to some moisture specification before weighing—many of the particles are larger and their actual mass concentration is greater in the air than is measured. For that reason alone, a given PM concentration would affect visibility more under humid conditions. There may be other factors in the relationship. The present discussion doesn't present the above issue, nor does it clarify the concept underlying the discussion. A couple of sentences would fix the problem.

CHAPTER 10

Most of the descriptions of PM over Mexico City were taken from white papers prepared by the MIT Integrated Program on Urban, Regional and Global Air Pollution. The two references cited should be corrected as suggested below (the order of the author list was wrong). It should also be pointed out that the white papers have been updated and are included as a chapter in the book edited by Molina and Molina (2002).

Molina, M.J., L. T. Molina, J. West, G. Sosa, and C. Sheinbaum Pardo, F. San Martini, M.A. Zavala and G. McRae (2002) "Air Pollution Science in the MCMA: Understanding Source-Receptor Relationships through Emissions Inventories, Measurements and Modeling," Chapter 5 in "*Air Quality in the Mexico Megacities: An Integrated Assessment,*" L.T. Molina and M.J. Molina, eds., Kluwer Academic Publishers, Dordrecht.

Page 10-18, line 5: It is egregious to say "careful coordination."

Page 10-29, line 11: "West et al. (2000a)" should be changed to "Molina et al. (2000)." See comment on page 10-35, line 31, below.

Page 10-31, line 16: "West et al. (2000b)" should be changed to "Sosa et al. (2000)." See comment on page 10-35, line 34, below

Page 10-35, line 31: The reference should be changed as follows (note the order of authors, and use the full name of the program, not "IPURGAP"):

Molina, M.J., L. T. Molina, J. West, F. San Martini, G. Sosa, and C. Sheinbaum (2000) "*Estado Actual del Conocimiento Científico de la Contaminación del Aire en el Valle de México*" (or "*Current State of Air Pollution Science in the Valley of Mexico.*" MIT-Integrated Program on Urban, Regional and Global Air Pollution Report No. 9, 84 pages.

Page 10-35, line 34: The reference should be changed as follows (note the order of authors, and use the full name of the program, not "IPURGAP"):

Sosa, G., J. West, F. San Martini, L. T. Molina and M. J. Molina (2000). *Air Quality Modeling and Data Analysis for Ozone and Particulates in Mexico City*. MIT- Integrated Program on Urban, Regional and Global Air Pollution Report No. 15, 76 pages.

CHAPTER 11

Page 11-1, line 12: Doesn't the standardization (or "harmonization") of CTMs fall under this broad theme?

Page 11-1, line 20: Does "detailed emission models" include better source emission characterization? They are different, but one can't have a better model without better characterization of source emissions.

Page 11-2, Science need 1.2: Does "carbon content" here mean speciation, quantity, or both? Be more explicit.

Page 11-2, Science need 2.2: Isn't it also important to assess trends in locations that have different characteristics? The present wording calls for collecting data over a longer time, which is good. Here or under a separate "need," it should call for developing a database by using comparable techniques in places that have different pollution and meteorologiccharacteristics.

Page 11-2, Science need 2.6: What does "harmonize" mean here? Without explanation, the word is ambiguous and therefore not useful. The same question applies to page 11-7, line 24; page 11-10, lines 12, 18, and 20; and page 11-16, line 20.

Page 11-2: Somewhere under the general heading of recommendation 2, the list of research needs should include better information on the surface chemistry of poorly soluble PM. Unless the PM is soluble, cells do not "care" about bulk chemistry, but about surface chemistry.

Page 11-3, Science need 3.3: It is not clear what is being sought by recommending "centers" or even what "centers" means. This item refers to Chapter 7.10, but the issue is not explained there either. The text on page 11-14 explains the recommendation adequately, but the wording in the table should be more explicit.

Page 11-3, Science need 4.3: It is not clear what "protocols" means. Does "protocols" mean (or include) models?

Page 11-3, Science need 4.9: It is not clear what is meant by "emissions processing interface."

Page 11-4, Science need 4.10: Why is this numbered "4.10" rather than "4.1"? What is meant by "self-consistent"?

Page 11-4, Science need 5.1: "Data is" should be "Data are."

Page 11-4, Science need 5.3: "Between" should be "Among."

Page 11-4, line 4: Change "understanding of properties" to "understanding of chemical and physical properties."

Page 11-5, line 2: Organic aerosols should be reduced because of their potential health effects and not just because they contribute to some extent to PM mass.

Page 11-6, line 23: A recent laboratory study (Bertram, A.K, A.V. Ivanov, M. Hunter, L.T. Molina and M. J. Molina. The reaction probability of OH on organic surfaces of tropospheric interest. *J. Phys. Chem.*, 105, 9415-9421, 2001) implies that particulate organic compounds become hygroscopic as they age (in any location).

Page 11-6, line 23: It seems unlikely that differences in the hygroscopicity of organic PM are due primarily to location, as the statement implies. It may vary among locations, but isn't it a function of the PM composition rather than location itself?

Page 11-7, line 6: Delete "and."

Page 11-7, line 11: Give examples of "artifacts."

Page 11-7, line 32: It should read "data are."

Page 11-8, line 3: It should read "are also needed."

Page 11-10, line 31: It is mentioned that Canada has an AWOS system, but it is not mentioned that the United States has both ASOS and AWOS systems.

Page 11-12, line 9: "then" should be "than."

Page 11-15, line 10: The committee does not agree that health effects typically depend on PM size. They are associated with PM size because composition is generally correlated with size. There are differences in deposition, and thus in location and dose, due to PM size, but there is so much overlap in size-related dosimetry in the different regions of the respiratory tract that PM composition is undoubtedly a bigger driver of health than PM size (within the respirable size range).

Page 11-21, lines 2, 4, and 6: "Data" is a plural word. It should be "Data are" and "They."

Page 11-21, line 12: "1000 a,b" should be "2000 a,b."

Page 11-22, lines 12-14: Certainly future assessments should extend to considerations beyond mass. This one does. Why is this rhetorical question even posed?

APPENDIX A

Page A6, line 6: Suggest reading CAM (2001) and Molina and Molina (2002) on the emission inventory of Mexico.